Lecture Notes in Physics

Volume 914

The Lecture Notes in Physics

The series Lecture Notes in Physics (LNP), founded in 1969, reports new developments in physics research and teaching-quickly and informally, but with a high quality and the explicit aim to summarize and communicate current knowledge in an accessible way. Books published in this series are conceived as bridging material between advanced graduate textbooks and the forefront of research and to serve three purposes:

- to be a compact and modern up-to-date source of reference on a well-defined topic
- to serve as an accessible introduction to the field to postgraduate students and nonspecialist researchers from related areas
- to be a source of advanced teaching material for specialized seminars, courses and schools

Both monographs and multi-author volumes will be considered for publication. Edited volumes should, however, consist of a very limited number of contributions only. Proceedings will not be considered for LNP.

Volumes published in LNP are disseminated both in print and in electronic formats, the electronic archive being available at springerlink.com. The series content is indexed, abstracted and referenced by many abstracting and information services, bibliographic networks, subscription agencies, library networks, and consortia.

Proposals should be sent to a member of the Editorial Board, or directly to the managing editor at Springer:

Christian Caron
Springer Heidelberg
Physics Editorial Department I
Tiergartenstrasse 17
69121 Heidelberg/Germany
christian.caron@springer.com

More information about this series at http://www.springer.com/series/5304

Jean-Pierre Rozelot • Coralie Neiner
Editors

Cartography of the Sun and the Stars

 Springer

Editors
Jean-Pierre Rozelot
Observatoire de la Côte d'Azur
Nice, France

Coralie Neiner
Observatoire de Paris-Meudon
Meudon, France

ISSN 0075-8450   ISSN 1616-6361   (electronic)
Lecture Notes in Physics
ISBN 978-3-319-24149-4   ISBN 978-3-319-24151-7   (eBook)
DOI 10.1007/978-3-319-24151-7

Library of Congress Control Number: 2016933870

Printed on acid-free paper

This Springer imprint is published by Springer Nature
The registered company is Springer International Publishing AG Switzerland

Preface

In the week of May 12–15, 2014, the Observatory of Besançon (France) welcomed a scientific community willing to develop new ways on how stars could be resolved and mapped to confront theory and observations. The cartography of the surface of the stars requires diverse skills, technique of analysis, and advanced modeling, i.e., the collaboration of scientists with various expertises. About 25 physicists and astronomers were able to debate, exchange, and share their knowledge in that rapidly developing field. Specific sessions were devoted to practical exercises, which encountered a real success. Following tradition, as this book is the fourth of a series, the speakers of this school were asked to supply a written version of their talks. Two additional chapters were added to provide a broader vision of the topic.

A particular attention has been paid to the Sun, with the invitation of solar experts in this area, because the Sun, due to its proximity, is a valuable laboratory for the mapping of all other stars. The knowledge gained on the Sun and the techniques developed are thus very important for scientists working on other stars.

Even in the best weather conditions, the instrumental diffraction limits drastically the angular resolution to perform astronomical imaging outside our solar system. Today, new techniques allow us for the first time to obtain nice images of stars. In particular, interferometry, combined with adaptive optics, recently allowed to reconstruct images of several stars. Already seven stars have been resolved in detail, in addition to the Sun of course.

This book takes stock of what was achieved with interferometry so far in Chile, on the ESO VLTI instrument and in the United States on the CHARA instrument. Physical aspects of the observations are important, especially in the case of rapidly rotating stars, for which the flatness and gravity darkening of the photosphere constrain models.

In addition to interferometric techniques, this book highlights mapping of surfaces of stars using Doppler or Zeeman–Doppler imaging methods, i.e., the use of spectroscopic or spectropolarimetric data to map spots, abundance, or magnetic fields at the stellar surface. It is also possible to resolve close binary stars by eclipse methods, which gives access to the interacting components.

This book also reports on the best images of the solar surface and connects the observable differential rotation to the underlying physical parameters. Recent measurements of flattening of the solar surface by SDO showed that its shape is linked to the rotation of its core. Such a result can probably be applied generally to stars.

The General Overlook of This Book Is as Follows. Chapter 1 by Aimé and Theys presents the basics of image reconstruction in astrophysics. Chapter 2 by Kosovichev and Zhao deals with the reconstruction of solar subsurfaces through local helioseismology from the GONG network and two space missions SOHO (Solar and Heliospheric Observatory) and SDO (Solar Dynamics Observatory). Chapter 3 by Lanza and Chap. 4 by Hiremath present results obtained from space photometry through helioseismology to map surface spots and thermal and magnetic field structures of the Sun. Chapter 5 by Rieutord shows how physical processes lead to the observed surface inhomogeneities. Chapters 6, 7, and 8, by Kervella, Perrin, and Domiciano de Souza, respectively, show the use of interferometric techniques to infer the shape, surface spots, and rotation of more distant stars. Finally, Chap. 9 by Kochukhov explains how spectroscopy and spectropolarimetry allow us to produce images of stars and, in particular, of their spots, abundance maps, and magnetic field configuration.

The authors wish to express their gratitude to all participants and speakers as the Besançon workshop permitted to anticipate the development of this particular branch of astrophysics, not only through future formal publications but also, and in many cases, through detailed discussions between specialists of different disciplines. The authors would also like to thank Jeff Kuhn, from the Institute of Astronomy of the Hawai University, for his plenary lecture, which could not be transcribed in this book.

We sincerely hope that all scientists, doctors, and students will be happy to find here the base of this new field of research, aimed at revealing the surface of stars.

Nice, France   Jean-Pierre Rozelot
Meudon, France   Coralie Neiner

Contents

Chapter 1
Reconstructing Images in Astrophysics, an Inverse Problem Point of View

Céline Theys and Claude Aime

Abstract After a short introduction, a first section provides a brief tutorial to the physics of image formation and its detection in the presence of noises. The rest of the chapter focuses on the resolution of the inverse problem. In the general form, the observed image is given by a Fredholm integral containing the object and the response of the instrument. Its inversion is formulated using a linear algebra. The discretized object and image of size $N \times N$ are stored in vectors $\mathbf{x}$ and $\mathbf{y}$ of length N^2. They are related one another by the linear relation $\mathbf{y} = H\mathbf{x}$, where H is a matrix of size $N^2 \times N^2$ that contains the elements of the instrument response. This matrix presents particular properties for a shift invariant point spread function for which the Fredholm integral is reduced to a convolution relation. The presence of noise complicates the resolution of the problem. It is shown that minimum variance unbiased solutions fail to give good results because H is badly conditioned, leading to the need of a regularized solution. Relative strength of regularization versus fidelity to the data is discussed and briefly illustrated on an example using L-curves. The origins and construction of iterative algorithms are explained, and illustrations are given for the algorithms ISRA, for a Gaussian additive noise, and Richardson–Lucy, for a pure photodetected image (Poisson statistics). In this latter case, the way the algorithm modifies the spatial frequencies of the reconstructed image is illustrated for a diluted array of apertures in space. Throughout the chapter, the inverse problem is formulated in matrix form for the general case of the Fredholm integral, while numerical illustrations are limited to the deconvolution case, allowing the use of discrete Fourier transforms, because of computer limitations.

C. Theys (✉) • C. Aime
UMR 7293 J.L. Lagrange, Observatoire de la Côte d'Azur, Université de Nice Sophia Antipolis, Parc Valrose, 06108 Nice Cedex, France
e-mail: celine.theys@unice.fr; claude.aime@unice.fr

© Springer International Publishing Switzerland 2016
J.-P. Rozelot, C. Neiner (eds.), *Cartography of the Sun and the Stars*,
Lecture Notes in Physics 914, DOI 10.1007/978-3-319-24151-7_1

1.1 Introduction

Due to diffraction of light, a point source on the sky produces, in the focal plane of the telescope, a response which is an extended pattern, called the point spread function (PSF). The observed focal plane image is obtained by substituting a weighted and shifted PSF to each point of the perfect geometrical image of the astronomical object. This operation is mathematically described by a two-dimensional Fredholm equation that simplifies to a convolution for a space invariant PSF. In this case the problem is conveniently treated in the Fourier space using the linear filtering of the optical transfer function (OTF). The resulting image is a blurred version of the object. The recorded image is contaminated by noises of several origins. A fundamental source of noise is the photodetection of the light, mathematically described by a Poisson transformation. Imperfections of detectors may add other sources of noises, generally assumed to follow Gaussian statistics. The problem of image reconstruction is to recover a result as close as possible to the object. We examine here, from an inverse problem point of view, a few techniques utilized for that.

The chapter is organized as follows. Basis of physical imaging are summarized in Sect. 1.2. The direct resolution of the inverse problem is discussed in Sect. 1.3, and the matrix formalism is introduced there. The principle of a regularization of the solution is described in Sect. 1.4. Iterative algorithms are explained in Sect. 1.5 on the examples of ISRA (Gaussian noise) and Richardson–Lucy (Poisson noise). An illustration for a diluted array of apertures is detailed in Sect. 1.6. Conclusions are given in Sect. 1.7.

1.2 Physical Basis of Optical Imaging

For a perfect circular telescope illuminated by a plane wave, the figure observed in the focal plane is an Airy function of angular size λ/D, where λ is the wavelength of the light and D is the diameter of the aperture. In presence of atmospheric turbulence, a large ground based telescope gives a random figure of speckles. Many elaborated techniques have been developed for their statistical exploitation for imaging (Labeyrie 1976; Roddier 1988; Weigelt 1991), but are beyond the scope of this work. We do not consider either conditions of long time exposure, studied by Fried (1966), because then the resolving power of the telescope is too drastically reduced, and is not very attractive for illustration of inverse problems. The kind of telescope response we use in this work corresponds to small defaults of the wavefront for a monolithic aperture. It may correspond to observations using a telescope in space, or using excellent adaptive optics for ground based observations.

1.2.1   Basic Relations for Image Formation

For a wave $\Psi_0(r)$ arriving from an on-axis point-source on the aperture of the telescope of transmission $P_0(r)$, we denote $P(r)$ the quantity:

$$P(r) = \Psi_0(r) P_0(r),\qquad(1.1)$$

where $r(r_x, r_y)$ is the position on the aperture. The complex-valued function $\Psi_0(r)$ expresses the defaults from a plane wave. In the focal plane, Fourier optics show that the observed intensity can be written as:

$$|\Psi_F(r)|^2 = \frac{1}{\lambda^2 F^2}|\hat{P}(\frac{r}{\lambda F})|^2,\qquad(1.2)$$

where F is the focal length of the telescope. The integral of $|\Psi_F(r)|^2$ over r gives the total flux passing through the telescope aperture. It is more convenient to use the angular units $\alpha(\alpha_x, \alpha_y)$, and to substitute to the true focal plane response a normalized quantity:

$$H(\alpha) = \frac{1}{\lambda^2 \mathbb{S}}|\hat{P}(\frac{\alpha}{\lambda})|^2\qquad(1.3)$$

where $\mathbb{S}$ is the telescope area. $H(\alpha)$ is indeed the function we take as the point spread function (PSF). It is normalized so that its integral over α is 1. An illustration of $P(r)$ and its corresponding PSF is given in Fig. 1.1. In this example, the defaults of the wavefront are small, and the corresponding PSF (right figure) is a slightly perturbed Airy spot. Now let us consider the incoherent imaging process. To derive the fundamental mathematical relations, we need to denote differently angular directions for the object and the image. We denote β the angular direction

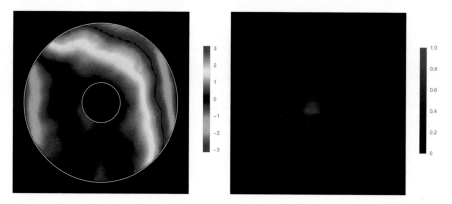

Fig. 1.1 *Left*: phase of the wavefront on a telescope aperture with a central obscuration. *Right*: corresponding PSF. The wavefront is the result of a numerical simulation that consists of filtering a white noise by a low-pass filter

corresponding to the object of intensity $X(\beta)$. An elementary two-dimensional angular region $d\beta$ of the object around the direction β irradiates an elementary intensity $X(\beta)d\beta$. This intensity is spread in the image plane as a function of the instrument response $H_\beta(\alpha)$. It produces the elementary intensity $dI(\alpha)$ of the form:

$$dI(\alpha) = X(\beta)d\beta \times H_\beta(\alpha). \tag{1.4}$$

For more generality, we have considered here the case in which the telescope response may vary in shape depending on the direction β of the source. This may be mandatory in the case of strong vignetting in an optical system, or if a large field of view is observed using an adaptive optics system optimized for a narrower zone.

Since the object is incoherent, the intensity in the focal plane is the sum of elementary intensities coming from all β directions. We have:

$$I(\alpha) = \int X(\beta)H_\beta(\alpha)d\beta \tag{1.5}$$

which is a Fredholm integral. Recover $X(\beta)$ from $I(\alpha)$ is the general problem that can be solved by the procedures described in this chapter, but this requires heavy numerical matrix computations, as developed in the next section.

The problem is much simpler if we can assume that the wavefront perturbation is independent of the direction of observation, i.e. that conditions of isoplanatism are fulfilled. Then the PSF is space invariant, and $H(\alpha + \beta)$ can be substituted to $H_\beta(\alpha)$ in the integral. The relation then simplifies to:

$$I(\alpha) = \int X(\beta)H(\alpha + \beta)d\beta = X(\alpha) * H(-\alpha), \tag{1.6}$$

which is the classical relation of convolution between the image, the object and the PSF. The minus sign comes from the inversion of the geometrical image by the lens. It is very often forgotten in the literature, but mandatory when the PSF is not an even function. An illustration of the smoothing effect of the PSF of Fig. 1.1 on a synthetic representation of an Earth-Moon system (NASA image) is shown in Fig. 1.2. Taking the Fourier transform of the terms of Eq. (1.6), we obtain:

$$\hat{I}(u) = \hat{X}(u).T(u) \tag{1.7}$$

where $u(u_x, u_y)$ are angular frequencies. $T(u)$ is the optical transfer function (OTF), Fourier transform of the PSF. This can be written as the aperture autocorrelation function:

$$T(u) = \mathcal{F}[H(-\alpha)] = \frac{1}{\mathbb{S}} \int P(r)P^*(r - \lambda u)dr \tag{1.8}$$

where $H(\alpha)$ is given by Eq. (1.3) and $P^*(r)$ is the complex conjugate of $P(r)$.

An illustration of $T(u)$, represented in modulus and phase, is given in Fig. 1.3. This function is zero for $u \geq D/\lambda$ for a circular aperture of diameter D, making the

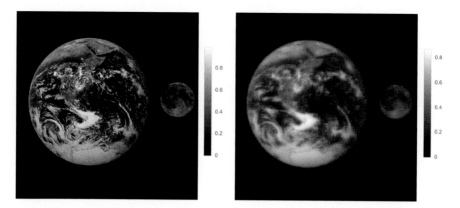

Fig. 1.2 *Left*: Synthetic object used for numerical illustration. *Right*: Noiseless focal plane image result of the convolution of the object by the PSF of Fig. 1.1

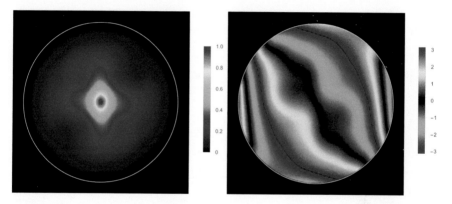

Fig. 1.3 OTF in modulus (*left*) and argument (*right*) corresponding to the PSF of Fig. 1.1. The *white circle* gives the aperture cut-off frequency

PSF a band limited function. Numerically, it is much easier to use the linear filtering of the OTF than to perform the convolution with the PSF. The focal plane noiseless image shown in Fig. 1.2 was indeed computed as the inverse Fourier transform of $\hat{I}(u)$ in the form:

$$I(\alpha) = X(\alpha) * H(-\alpha) = \mathcal{F}^{-1}[\hat{X}(u).T(u)]. \tag{1.9}$$

1.2.2   Image Detection in the Presence of Noises

The quantity $I(\alpha)$ gives the intensity of the image in the classical sense. The first source of noise comes with the fundamental process of photodetection. A sensor such as a CCD collects the light received inside elementary surfaces that further lead to pixels in the image. For an integrated energy of m photons (a real positive number), the effective number of photons detected is a random integer variable n following the Poisson law $P(n/m) = \exp(-m)m^n/n!$, as described in Goodman (2012) for example.

There are other sources of noises. For example, the detector is read by an electronic process which adds a noise independent of the number of photoelectrons in a pixel. It is usually assumed a Gaussian law for that, of the form $\mathcal{N}(g, \sigma^2)$, of mean g and variance σ^2.

For the sake of clarity and easiness of reading, we will consider only two extreme cases. The first one considers that the read-out noise is negligible, taking only into account the Poisson transformation:

$$Y(\alpha) \sim \mathcal{P}(I(\alpha)). \tag{1.10}$$

Figure 1.4, top curves, gives an illustration for a total number of 30×10^6 photons in the whole image. Curves have been normalized for comparison with noiseless

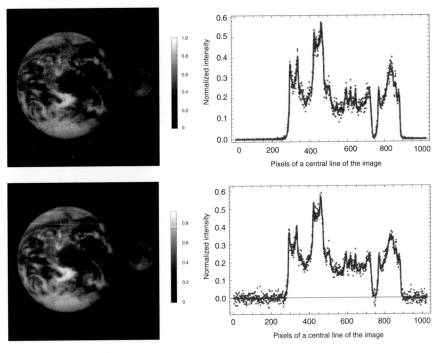

Fig. 1.4 *Top left*: photodetected image with 30 million photons (level 0.2 corresponds to 100 photons per pixel). *Right*: horizontal center cut. *Bottom*: *Left*: Image with Gaussian noise $\sim \mathcal{N}(0, 0.02)$. *Right*: horizontal center cut

data, and the levels are no more integer numbers. A typical value for a pixel of the Earth-like image is 100 photoelectrons.

The second model assumes that the number of photons is sufficiently large to neglect the Poisson transformation and consider only the read-out noise, modeled as a signal-independent additive noise:

$$Y(\alpha) = I(\alpha) + B(\alpha). \tag{1.11}$$

Figure 1.4, bottom curves, is an illustration of Eq. (1.11), with $B(\alpha)$ that follows a zero-mean Gaussian law $\mathcal{N}(0, 0.02)$. That standard deviation roughly gives a level of noise comparable to that of the Poisson noise, where the image is non-zero. Differences between the two models appear clearly in the regions where $I(\alpha) \simeq 0$, since $P(n/0) = 0$. Arrays used in numerical simulations are 1024×1024 pixels, but only the central part of the images is represented (zoom $\times 5$).

1.3 Inverse Imaging Problem

For the purposes of calculation it is necessary to use discretized quantities. Thus the observed process described in Eq. (1.11) can be written as:

$$y(m, n) = \sum_{i,j} x(i, j) h(m, i, n, j) + b(m, n) \tag{1.12}$$

for the general case of the Fredholm equation of Eq. (1.5), which reduces to the discrete form of the convolution of Eq. (1.6):

$$y(m, n) = \sum_{i,j} x(i, j) h(m - i, n - j) + b(m, n) \tag{1.13}$$

where y, h, x and b are numerical values of tables (indexed by m, n) corresponding to discretized quantities Y, H, X and B, respectively.

1.3.1 The Raw Inverse Filter and an Improved Version

Considering the space-invariant case of Eq. (1.13), an immediate naive solution to estimate the object from the data is to take the Discrete Fourier Transforms (DFT) of y and h and to recover the DFT of the object by a simple division:

$$\hat{x}(k, l) = \frac{\hat{y}(k, l)}{\hat{h}(k, l)} \tag{1.14}$$

where (k, l) have generally the same dimensions than (m, n) and $\hat{x}$ is the DFT of x. The estimated object is then taken as the inverse DFT of $\hat{x}(k, l)$. The division can be made only where $\hat{h}(k, l)$ is different from zero. A simple treatment of that is to limit the domain of the division to the region where the OTF is non-zero, setting $\hat{x}(k, l) = 0$ where $\hat{h}(k, l) = 0$ or very small. For that we can insert a multiplicative window function $F(k, l)$ in Eq. (1.14):

$$\hat{x}(k, l) = F(k, l) \times \left(\frac{\hat{h}(k, l)\hat{x}(k, l)}{\hat{h}(k, l)} + \frac{\hat{b}(k, l)}{\hat{h}(k, l)} \right), \qquad (1.15)$$

where we have replaced $\hat{y}$ by the direct transformation in Eq. (1.13).

In the example of Fig. 1.5, we give the result obtained for the photodetected image, limiting the division in Eq. (1.15) to the spectral bandwidth at 98 % of the telescope frequency cut-off. The representation of the corresponding inverse filter is clipped to twenty in the figure, while true values reach several hundred close to the cut-off frequency. Trying to use 99 % of the cut-off frequency would induce a correcting factor of the order of one thousand.

An intuitive improvement would be to use a smooth function for F, depending on the signal to noise ratio. This is the effect of the Wiener filter, developed for Gaussian noise. It is equal to the ratio of the object power spectrum to the sum of the object power spectrum and the noise power spectrum. Some a priori knowledge of these quantities is necessary. Nevertheless, to implement this correction, we have used for F the ideal theoretical OTF of the telescope that is a damping function for high frequencies. This has given excellent results in this case. Note that the problem of the inversion of H exists also for a noiseless image, which is characteristic of an ill posed inverse problem. In the following, we develop a matrix formulation of the model to search for better solutions.

1.3.2 Matrix Formulation

The discretized form of the Fredholm integral given in Eq. (1.12) can be written in the following matrix form:

$$\mathbf{y} = H\mathbf{x} + \mathbf{b} \qquad (1.16)$$

where $\mathbf{y}$, $\mathbf{x}$ and $\mathbf{b}$ are vectors of the discretized values of the image, the object and the noise lexicographically stored. Lexicographic means that the values of each table read from left to right and from top to bottom are arranged in a vector. An image of size $N \times N$ then yields a vector of length N^2. Consequently the matrix H is of dimensions $N^2 \times N^2$. Each line of H sums to 1 since each line contains all the coefficients of the normalized PSF. We denote in bold the vectors and in capital letters the matrices. A realization of the value of the image in the pixel i is written

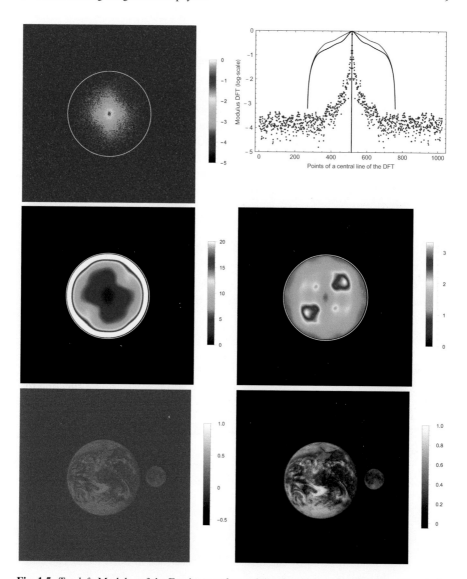

Fig. 1.5 *Top-left*: Modulus of the Fourier transform of the photodetected image. The *white circle* is the telescope frequency cut-off. *Top-right*: central cut of the *left figure*, with the telescope theoretical OTF (in *red*) and the modulus of $T(u)$, all *curves* in log-scale. *Middle-left*: modulus of the inverse filter saturated at 20. *Middle-right*: modulus of the inverse filter "regularized" by the telescope theoretical OTF. *Bottom-left*: image given by the inverse filter. *Bottom-right*: image given by the "regularized" inverse filter

in the following matrix form:

$$y_i = (H\mathbf{x})_i + b_i. \tag{1.17}$$

Considering the space-invariant case of Eq. (1.13), the matrix H is block Toeplitz and each block is Toeplitz, so H has a Toeplitz-block-Toeplitz structure. In this case, the matricial product $H\mathbf{x}$ can be advantageously performed in the Fourier space using two-dimensional DFT on the $N \times N$ arrays. Nevertheless we keep the matricial notation for more generality.

The problem without constraints on the solution is then to restore $\mathbf{x}$ from the data $\mathbf{y}$ with the knowledge or the measure of the PSF, H.

1.3.3 Minimum Variance Unbiased Solutions

From the model described in Eq. (1.16), without noise or with unknown noise, a classical estimate is the solution that minimizes the quadratic error:

$$J_1(\mathbf{x}|\mathbf{y}) = ||\mathbf{y} - H\mathbf{x}||^2. \tag{1.18}$$

This term is called the fidelity term in the sense of fidelity with respect to the measurements, and the argument of the minimum of J_1 is called the least squares solution, $\mathbf{x}_{LS}$:

$$\mathbf{x}_{LS} = arg\min_{\mathbf{x}} J_1(\mathbf{x}) = (H^T H)^{-1} H^T \mathbf{y} = H^\dagger \mathbf{y} \tag{1.19}$$

where $H^\dagger$ is called the generalized inverse matrix and $\mathbf{x}_{LS}$ is also called the generalized inverse solution. If the matrix H is square then the solution reduces to:

$$\mathbf{x}_{IF} = H^{-1} \mathbf{y}, \tag{1.20}$$

which is the solution given by the inverse filter described in Sect. 1.2.1. The same solution is obtained if noise components are independent and identically distributed (i.i.d).

If the noise has a known autocorrelation matrix, then the solution is called the Generalized Least Squares solution, $\mathbf{x}_{GLS}$:

$$\mathbf{x}_{GLS} = (H^T R^{-1} H)^{-1} H^T R^{-1} \mathbf{y}, \qquad R = E[\mathbf{b}\mathbf{b}^T]. \tag{1.21}$$

It is important to notice that from a statistical point of view, all the previous solutions, $\mathbf{x}_{LS}$, $\mathbf{x}_{IF}$ and $\mathbf{x}_{GLS}$ are the best solutions in the sense of unbiased estimators and of minimum variance (Kay 1993).

In conclusion, in the discrete case, although a solution can always be obtained, the inversion of H or $H^T H$ (if H is not square) is a difficult operation because H is badly conditioned.

A numerical solution consists of increasing the conditioning of H by suppressing the smaller singular values of the singular decomposition of H and by computing the inverse of H by:

$$(H^T H)^{-1} = \sum_{i=1}^{\mathbf{K}} \frac{1}{\lambda_i^2} \mathbf{v}_i \mathbf{v}_i^T \tag{1.22}$$

where λ_i are the singular values of H or H^T arranged in the decreasing order and $\mathbf{v}_i$ are the corresponding singular vectors, K is the order of the truncation. This solution has the main drawback of the choice of the truncation order K that moreover cannot be linked to physical considerations.

1.4 Adding A Priori: Regularization

The "best" solution obtained in Sect. 1.3 is not a good solution in the context of ill-posed problems. A classical way to stabilize the solution is to regularize the problem. From a statistical point of view, the estimator is biased and no result on the variance of the estimators is available. In other words, only empirical considerations or numerical computations can be made to evaluate the quality of the estimation.

A classical way to regularize the problem is to add a regularization term J_2 to the fidelity term J_1 and to minimize a composite criterion J:

$$J(\mathbf{x}|\mathbf{y}) = J_1(\mathbf{x}|\mathbf{y}) + \gamma J_2(\mathbf{x}). \tag{1.23}$$

The regularization coefficient γ tunes the weight of the fidelity term versus the weight of the regularization term. The choice of J_1 and J_2 is qualitative, J_1 is generally assigned by the data model while J_2 is chosen to impose a smoothness constraint on the solution. A well known class for regularization terms J_2 is the Tikhonov regularization (Tikhonov 1976). The regularization considered in the following is a Tikhonov regularization.

In the linear form of Eq. (1.16) with a quadratic error for J_1 as in Eq. (1.18), an interesting choice for J_2 is the quadratic one:

$$J_2(\mathbf{x}) = ||\mathbf{x} - \bar{\mathbf{x}}||^2. \tag{1.24}$$

By minimizing this term, we minimize the Euclidean distance between **x** and a default solution $\bar{\mathbf{x}}$, that must be chosen "smooth" in order to pull the solution towards a "smooth" object. Then the composite criterion, Eq. (1.23) is:

$$J(\mathbf{x}) = ||\mathbf{y} - H\mathbf{x}||^2 + \gamma ||\mathbf{x} - \bar{\mathbf{x}}||^2 \qquad (1.25)$$

and in this case the minimizer is explicit:

$$\mathbf{x}_{RLS} = arg \min_{\mathbf{x}} J(\mathbf{x}) = (H^T H + \gamma)^{-1}(H^T \mathbf{y} + \gamma \bar{\mathbf{x}}). \qquad (1.26)$$

We find the non-regularized solution, Eq. (1.19), by setting $\gamma = 0$ and on the contrary $\mathbf{x}_{RGLS} \rightarrow \bar{\mathbf{x}}$ for $\gamma = \infty$.

The choice of $\bar{\mathbf{x}}$ is both important and limited. If no information about the object is available, taking any informative $\bar{\mathbf{x}}$ a bit different from the true object pulls the result towards this false object. In this case, the most non informative choice, allowing to stabilize the solution must be taken, so that the choice of a constant value for $\bar{\mathbf{x}}$ appears to be a satisfactory default solution.

In the particular quadratic case of Eq. (1.25), the tuning of the regularization parameter can be made by the method of the L-curve (Hansen 1992). The L-curve is a plot of the regularization term J_2 versus the fidelity data term J_1. The resulting curve has the shape of an L and the value of γ at the angle of the L corresponds to a trade-off between the error due to the regularization and the error due to the fidelity term. It is the value suggested by this approach.

The regularized least squares estimator of Eq. (1.26), has been computed for the image with Gaussian noise, and the results are given in Fig. 1.6. In practice, all the quantities have been computed by DFT, products of DFTs and inverse DFT of the arrays of the PSF, the OTF and the image. The regularization parameter has been tuned by the method of the L-curve and the corresponding "best" regularized solution is obtained for $\gamma = 0.008$, a value very close to the one that could be obtained by comparison with the true object. In Fig. 1.6, under-regularized ($\gamma = 2 \times 10^{-5}$) and over-regularized ($\gamma = 0.2$) solutions are shown. Adding a regularization term clearly improves the result, the regularized solution must be compared to the one obtained by the inverse filter, Fig. 1.5.

Some considerations can be made from the numerical experiment of Fig. 1.6. The L-curve is a good non supervised method for choosing the regularization parameter. The dynamic of the color bars gives information on the behavior of the algorithm. A first point is that the reconstructed image can have negative values since no positivity constraint on the solution is imposed. If the coefficient of regularization is chosen too low, the image is not sufficiently stabilized and the amplitude of high frequency increases. On the contrary, if the coefficient of regularization is chosen too large, the estimated solution will tend to the default solution, 1 in this example. In this case there is no negative value in the image. The two main limitations of this estimator are that the non-negativity constraint cannot be imposed and that

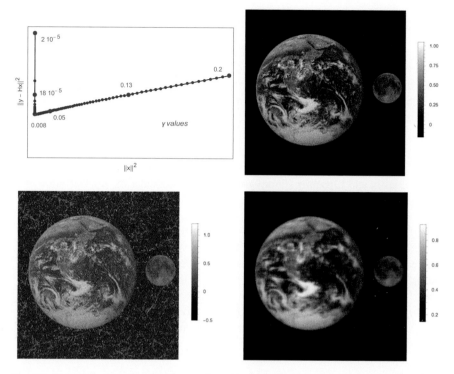

Fig. 1.6 *Top left*: L curve for different γ values. Central part of the images (zoom ×5) obtained for: $\gamma = 0.008$ (*top right*), $\gamma = 2 \times 10^{-5}$ (*bottom left*) and $\gamma = 0.2$ (*bottom right*). Note the negative parts in the image for the weak regularization parameter

the analytical solution exists only in the case of the linear quadratic model of Eq. (1.25).

In astrophysics, a classical model is the Poisson model of Eq. (1.10). In this case, the model is not linear anymore and there is not an analytical solution. We have to turn towards iterative methods that allow the introduction of a non-negativity constraint and make it possible to stop the iterations before convergence. Effectively, at convergence, the "optimal" solution is reached but, as we have seen, this solution is a bad one for ill-posed problems.

1.5 Iterative Methods and Non-negativity Constraint

The use of iterative methods is justified for one or more of the following problems:

- The problem is large and computation of H^{-1} is too expensive in computation time.
- There is no analytical solution for the model under consideration.

- The introduction of inequality constraints makes the problem non-linear with respect to unknowns.

In these cases, we search for the solution by minimizing iteratively the criterion J with eventual constraints.

In the following, we present a succinct overview of optimization methods just enough to retrieve the two well-known algorithms Richardson Lucy and ISRA. The problem is to solve:

$$\min_{\mathbf{x}} J(\mathbf{x}), \quad s.t \quad x_i \geq 0, \quad \forall i \tag{1.27}$$

with J convex and Lipschitz (finite) gradient. If we consider the minimization of J without constraints, a classical iterative method is the gradient descent one:

$$\mathbf{x}^{k+1} = \mathbf{x}^k + \alpha^k[-\nabla_{\mathbf{x}} J(\mathbf{x}^k)], \tag{1.28}$$

where $\mathbf{x}^{k+1}$ is the iterate $k+1$, $\nabla_{\mathbf{x}} J$ is the gradient of J with respect to (w.r.t) $\mathbf{x}$ and α^k is the descent step size, tuned to ensure the descent of the algorithm, i.e. $J(\mathbf{x}^{k+1}) \leq J(\mathbf{x}^k)$.

Minimizing a convex cost function J under inequality constraints can be classically achieved by introducing the Lagrange function $\mathcal{L}$ associated to the problem without constraints that is for the non-negativity constraint, Eq. (1.27):

$$\mathcal{L}(\mathbf{x}, \boldsymbol{\lambda}) = J(\mathbf{x}) - \boldsymbol{\lambda}^T \mathbf{g}(\mathbf{x}),$$

where $\boldsymbol{\lambda}$ is the vector of Lagrange multipliers and $\mathbf{g}(\mathbf{x})$ is the vector of a function of $\mathbf{x}$, that must be chosen to express the non-negativity constraints. The Karush Kuhn Tucker (KKT) conditions (Karush 1939; Kuhn and Tucker 1951) at the optimum $(\mathbf{x}^\star, \boldsymbol{\lambda}^\star)$ express as follows

$$[\nabla_{\mathbf{x}} \mathcal{L}(\mathbf{x}^\star, \boldsymbol{\lambda}^\star)]_r = 0, \qquad \forall r, \tag{1.29}$$

$$g(\mathbf{x}_r^\star) \geq 0, \qquad \forall r, \tag{1.30}$$

$$\lambda_r^\star \geq 0, \qquad \forall r, \tag{1.31}$$

$$\lambda_r^\star \, g(\mathbf{x}_r^\star) = 0, \qquad \forall r, \tag{1.32}$$

where $\nabla_{\mathbf{x}} \mathcal{L}$ is the gradient of $\mathcal{L}$ w.r.t. $\mathbf{x}$ and the notation $[\cdot]_r$ is used for the rth component of a vector. Once the system (1.29) is solved, the KKT conditions are reduced to the following condition:

$$[\nabla_{\mathbf{x}} J(\mathbf{x}^\star)]_r \, g(\mathbf{x}_r^\star) = 0, \qquad r = 1, \ldots, R. \tag{1.33}$$

R is the number of components. This equation gives the condition that must be satisfied at the optimum of the function J under the constraint of non-negativity on the parameters $\mathbf{x}$. The simplest choice for $g(\cdot)$ is $g(\mathbf{x}_r) = \mathbf{x}_r$. Then, more generally, if we can ensure that $\mathbf{x}_r$ remain positive or zero along the iterations, we can propose a modified gradient descent algorithm:

$$x_r^{k+1} = x_r^k + \alpha_{r}^k f_r(x^k) x_r^k [-\nabla_x J(x^k)]_r. \tag{1.34}$$

where $f_r(\mathbf{x})$ is a positive function. In this equation, $\alpha_r^{(k)}$ is the descent step-size that must be adjusted to ensure the non-negativity of the estimate and the convergence of the algorithm.

The iterative form of Eq. (1.34) is a general gradient descent algorithm that solves the problem of Eq. (1.27) for an appropriate chosen value of α_r^k. In the following, we consider the two functionals J corresponding to the poissonian model, Eq. (1.10) and the Gaussian model, Eq. (1.11) and a particular choice of f and α will lead to the so-called Richardson Lucy and ISRA algorithms.

1.5.1 Poisson Process, Richardson Lucy Algorithm

In this part, the considered model is a discretization of the Poisson model of the data, and Eq. (1.10) becomes for a pixel i:

$$y_i = \mathcal{P}\left((H\mathbf{x})_i\right). \tag{1.35}$$

Since pixels are independent w.r.t. the Poisson process, the probability density for N pixels is:

$$P(\mathbf{y}|\mathbf{x}) = \prod_{i=1}^{N} \frac{(H\mathbf{x})_i^{y_i}}{y_i!} \exp\left(-(H\mathbf{x})_i\right). \tag{1.36}$$

Maximizing the log-likelihood $\log(P(\mathbf{y}|\mathbf{x})$ is equivalent to minimize the objective function $J(\mathbf{x}) = -\log(P(\mathbf{y}|\mathbf{x}))$ that is, to an additive constant:

$$J(\mathbf{x}) = \sum_{i=1}^{N}((H\mathbf{x})_i - y_i \log(H\mathbf{x})_i). \tag{1.37}$$

The argument of the minimum of J is not explicit, and using an iterative algorithm is then mandatory. The gradient is easily computed and can be written in the following matrix form:

$$\nabla_x J(\mathbf{x}) = H^T(-\mathbf{y}./H\mathbf{x} + 1_N) \tag{1.38}$$

where . denotes the Hadamard product, i.e the product term by term and $1_N =$ $(11\ldots1)^T$. Then if we take the iterative form, Eq. (1.34) and setting $\alpha_r^k = 1$, $\forall r$, $\forall k$, we obtain the following multiplicative form called the Richardson–Lucy (RL) algorithm (Richardson 1972; Lucy 1974, 1994):

$$x_r^{k+1} = x_r^k[H^T(\mathbf{y}./(H\mathbf{x}^k))]_r \qquad (1.39)$$

where we have made use of the property $H^T 1_N = 1$. The convergence of Eq. (1.39) has been demonstrated. This algorithm, like all multiplicative algorithms, ensures the non-negativity of x^k for a non-negative initial value x^0. Moreover it has the property of conserving the flux i.e $\sum_r x_r^{k+1} = \sum_r x_r^k$. These properties have made it very popular in astrophysics.

As the deconvolution problem is an ill-posed problem, instability in the solution appears as the number of iterations increases. The problem is then to stop them to get a physically satisfactory solution and for that, to determine the optimal iteration number (Lucy 1994).

The RL algorithm of Eq. (1.39) has been applied on the image with Poisson noise, Fig. 1.4 top left and results are shown in Fig. 1.7. Figure 1.7 left is the best reconstructed object obtained by stopping the process at the optimal number of iterations, taken as the one that minimizes the normalized quadratic error:

$$k_{opt} = arg\min_k \frac{(\mathbf{x}^k - \mathbf{x})^T(\mathbf{x}^k - \mathbf{x})}{\mathbf{x}^T\mathbf{x}}. \qquad (1.40)$$

Figure 1.7 right is the reconstructed object at iteration 500, the result deteriorates with the increasing number of iterations. Let us note that in real experiments, the optimal iteration cannot be computed.

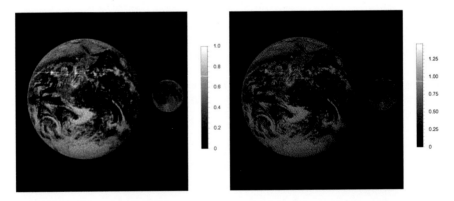

Fig. 1.7 Illustration of the Richardson–Lucy algorithm. *Left*: Best reconstructed object ($k_{opt} = 36$). *Right*: Reconstructed object at iteration $k = 500$ showing the noise amplification

1.5.2 Gaussian Process, ISRA Algorithm

In this part, the considered model is the Gaussian one of Eq. (1.17). The approach to achieve the algorithm is substantially the same as for the RL algorithm. Since the noise b is i.i.d, the probability density for N pixels is:

$$P(\mathbf{y}|\mathbf{x}) = (\sigma\sqrt{2\pi})^{-N} \prod_{i=1}^{N} \exp\left(-\frac{(y_i - (H\mathbf{x})_i)^2}{2\sigma^2}\right) \tag{1.41}$$

and the objective function $J(\mathbf{x})$ is naturally the same as in Eq. (1.18). As we saw it in Sect. 1.3.3, the usual minimizers of J are unacceptable in the context of inverse ill-posed problems and moreover in physical applications it is necessary to add a constraint of non-negativity. The gradient of J is:

$$\nabla_{\mathbf{x}} J(\mathbf{x}) = H^T H\mathbf{x} - H^T \mathbf{y}. \tag{1.42}$$

In the same way that for the RL algorithm, we take the iterative form of Eq. (1.34) with $\alpha_r = 1$ and $f_r(x^k) = 1/[H^T H\mathbf{x}^{(k)}]_r$. We then obtain the multiplicative form of the modified gradient descent algorithm, called the iterative space reconstruction algorithm (ISRA) by Daube-Witherspoon and Muehllehner (1986):

$$x_r^{(k+1)} = x_r^{(k)} \frac{[H^T \mathbf{y}]_r}{[H^T H\mathbf{x}^{(k)}]_r}. \tag{1.43}$$

The convergence of Eq. (1.43) has been demonstrated in DePierro (1987). An important point is that this algorithm does not ensure the non-negativity of $\mathbf{x}$ for values of $\mathbf{y}$ negative.

ISRA, Eq. (1.43), has been applied on the image with Gaussian noise, Fig. 1.4 bottom left and results are shown in Fig. 1.8 . Figure 1.8 left is the best reconstructed object obtained by stopping the process at the optimal number of iterations, taken as the one that minimizes the normalized quadratic error, Eq. (1.40). Figure 1.8 right is the reconstructed object at iteration 100 showing the noise amplification with the increasing number of iterations. Note that the reconstructed objects have negative values in both cases, as indicated in the color bars, and the noise amplification is particularly dramatic in the region near the moon, completely altering the image.

Fig. 1.8 Illustration of ISRA. *Left*: Best reconstructed object (k_{opt} = 19). *Right*: Reconstructed object at iteration k = 100 showing the noise amplification

1.6 Illustration of RL Algorithm for Observations Made with a Diluted Array of Telescopes

For the monolithic aperture considered in the previous section, the OTF is a low pass filter with a well defined cut-off angular frequency $|u_c| = D/\lambda$. The first goal of an iterative algorithm such as RL is to retrieve the best as possible the spectral information in this domain, and try to recover some information coming from outside the cut-off frequency. For a diluted array of apertures, the problem is more complex. Then the OTF stands in several non-continuous regions, according to the number of apertures and the configuration.

The problem of the optimal configuration for a given number of apertures is a very stimulating mathematical problem, as described by Kopilovich and Sodin (2001), for example. The configuration we have used in our simulation (see Fig. 1.9, top left), is made of $M = 35$ identical apertures set on a regular two-dimensional grid in a non-redundant configuration, which means so that there is never two identical differences of positions between apertures. We do not claim that this configuration is the optimal one for 35 non redundant apertures, but it presents interesting features to illustrate the behavior of the RL algorithm. We assume that all apertures can form an image at a common focus, the array being perfectly coherent with no phase aberration. Such an experiment is very far from being feasible now, but might be possible in the future for a space-born array of telescopes.

The OTF of this configuration fills fairly regularly the low frequencies, notwithstanding void spaces between the regions due to distances between elementary apertures. For the sake of clarity, only the positions of the regions are represented in Fig. 1.9, top right. In fact, the OTF of an elementary aperture should be substituted to each point in the figure. This result is directly obtained from the relations we have given in Sect. 1.2. Heuristically one may come to the same result just considering that any two given apertures form an elementary stellar Michelson interferometer.

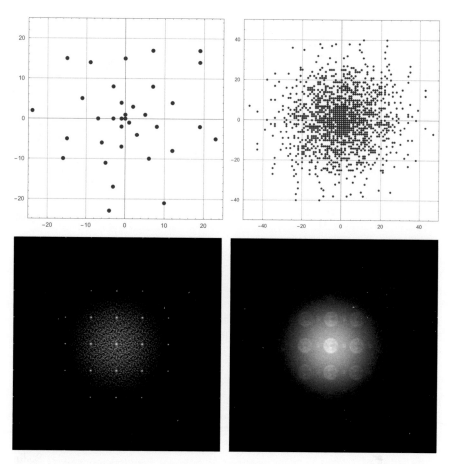

Fig. 1.9 Aperture configuration of the 35 diluted apertures set in a non redundant configuration (*top left*), corresponding MTF (*top right*) and PSF (*bottom left*). Focal plane noiseless image (*bottom right*)

Setting as before the value of the OTF to 1 for $u = 0$, the level of each elementary OTF is $1/2M$, as it is 1/4 for the Michelson stellar interferometer.

For our simulation done using 1024×1024 points, the elementary distance between two apertures is 8 points while the diameter of the telescope is 3 points. In Fig. 1.9 the array is presented in a box of 50×50 elementary distances, and the OTF in a box of 100×100 elementary distances, i.e. 800×800 points.

The corresponding PSF (bottom left of Fig. 1.9) possesses two features. First an Airy-like response is replicated on a two-dimensional grid. In a first approximation, this Airy pattern is that of the giant meta-telescope we seek to synthesize, and the replication step is the inverse of the elementary distance between telescopes. A random speckle-like structure appears added to this regular pattern. It reflects the fact that the configuration of 35 telescopes does not form a perfect regular

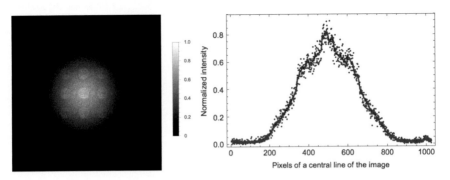

Fig. 1.10 Noisy focal plane image of the 35 apertures NR array for 30 M photons in the image. The level 0.6 corresponds to 200 photons per pixel in the horizontal center cut. *Red curve*: noiseless image

grid that would otherwise require 2500 elementary apertures. The corresponding noiseless focal plane image is given in Fig. 1.9, bottom right. The object is replicated according to peaks in the PSF. To keep the illustration possible for a representation of 1024×1024 pixels, the Earth-Moon object has been slightly reduced in resolution, and the relative distance between the Earth and the Moon increased such as to introduce a visual ambiguity in the resulting image, whether the Moon is at the right or at the left of the Earth.

The photodetected image is given in Fig. 1.10 and corresponds also to 30×10^6 photons in the image, as in the former simulation for the monolithic aperture. An horizontal cut of the image is provided in the figure and compared with the corresponding noiseless representation. This cut makes it visible the high level of the diffuse image produced by the speckle-like part of the PSF.

An illustration of the results obtained at different iteration numbers ($k = 5$, 100 and the best reconstruction obtained for 3866) of the RL algorithm is given in Fig. 1.11, left curves, together with the original object. As k is increased, the dominant visual effect on the reconstructed object x^k is the reduction of the number of replicas towards a single object-like Earth-Moon image. It is interesting to see the evolution of the algorithm in the Fourier plane, as represented in Fig. 1.11, right curves. The suppression of replicas corresponds to a filling of the frequency gaps in the OTF of the NR array. This phenomenon was described in Aime et al. (2012) and in Mary et al. (2013). During the iterations there is a progressive filling of the Fourier plane similar to the expansion of an ink stain on a blotting paper, resulting to a progressive in painting from known to unknown regions of the Fourier plane. We have no rigorous mathematical explanation for this effect that deserves further studies beyond the limits of this work.

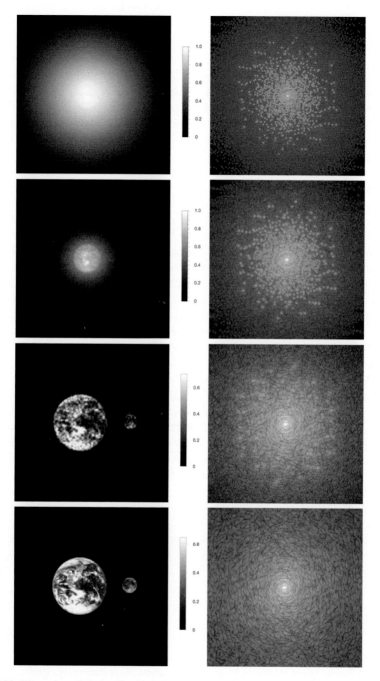

Fig. 1.11 Illustration of the application of the RL algorithm to the $30\,M$ photons image of the NR array of 35 apertures of Fig. 1.10. *Left curves*: direct plane, *right curves* modulus of the corresponding Fourier transform in log scale. *From top to bottom*, results of RL at iterations 5 and 100 (zoom $\times 2.5$), best reconstruction at iteration $k = 3866$ (*left*) and original object (*right*) (zoom $\times 5$). See the text for comments on image spectra

1.7   Concluding Remarks

This chapter provides elements of methodology for solving an ill-posed inverse problem. We have seen that for such problems an optimal solution in the statistical sense is irrelevant. The solution must be stabilized by regularizing the problem or by using an iterative algorithm and stopping the iterations. The problem then becomes finding the optimal number of iterations. Throughout the chapter numerical illustrations are proposed for an extended object observed using monolithic or diluted apertures.

The study focuses on two basic models, the presence of an additive Gaussian noise and the Poisson transformation for a photodetected image. For the Gaussian model, there is an analytical regularized solution that can be qualified of unsupervised in the sense that a correct value of the regularization coefficient can be obtained by the method of the L-curve, for example. However, this solution does not ensure the non-negativity of the solution. For the Poisson model, there is no analytical solution, then we turn to iterative methods such as the Richardson Lucy algorithm. This multiplicative algorithm minimizes the Kullback–Leibler divergence, associated to the Poisson model and guarantees the non-negativity of the solution and the conservation of the flux. For the Gaussian model, an iterative algorithm may be necessary in the case where the inversion of H is too expensive in computation time. A well-known algorithm is ISRA but its use in its original version does not guarantee the non-negativity of the solution.

The presentation of inverse problems for reconstructing images in astrophysics we have given in this chapter must be considered as an introduction to the subject. For example, the problem of stopping the iterations can be solved by using a regularized algorithm which stabilizes the error when the number of iterations increases (Titterington 1985; Demoment 1989; Bertero et al. 1995). More sophisticated models have been developed for reconstruction of images corrupted by Poisson and Gaussian noise (Llacer and Nunez 1990; Nuñez and Llacer 1993, 1998; Wu and Anderson 1997; Lantéri and Theys 2005; Benvenuto et al. 2008; Snyder et al. 1993, 1995).

Acknowledgements We are grateful to organizers of the Besançon school for their invitation and to H. Lantéri for very fruitful discussions.

References

Aime, C., Lantéri, H., Diet, M., & Carlotti, A. (2012). Strategies for the deconvolution of hypertelescope images. *Astronomy and Astrophysics, 543,* A42.

Benvenuto, F., La Camera, A., Theys, C., Ferrari, A., Lantéri, H., & Bertero, M. (2008). The study of an iterative method for the reconstruction of images corrupted by Poisson and Gaussian noise. *Inverse Problems, 24,* 035016.

Bertero, M., Boccaci, P., & Maggio, F. (1995). Regularization methods in image restoration: An application to HST images. *International Journal of Imaging Systems and Technology, 6,* 376–386.

Daube-Witherspoon, M. E., & Muehllehner, G. (1986). An iterative image space reconstruction algorithm suitable for volume ECT. *IEEE Transactions on Medical Imaging, 5,* 61–66.

Demoment, G. (1989). Image reconstruction and restoration: Overview of common estimation structures and problems. *IEEE Transactions on ASSP, 12,* 2024–2036.

DePierro, A. R. (1987). On the convergence of the image space reconstruction algorithm for volume RCT. *IEEE Transactions on Medical Imaging, 2,* 328–333.

Fried, D. L. (1966). Optical resolution through a randomly inhomogeneous medium for very long and very short exposures. *Journal of the Optical Society of America, 56,* 1372.

Goodman, J. W. (1985). *Statistical optics.* New York: Wiley.

Hansen, P. (1992). Analysis of discrete ill-posed problems by means of the L-curve. *SIAM Review, 34,* 561.

Karush, W. (1939). Minima of functions of several variables with inequalities as side constraints. Ph.D., University of Chicago.

Kay, S. M. (1993). Fundamentals of statistical signal processing: Detection theory. Prentice-Hall, Inc.

Kopilovich, L. E., & Sodin, L. G. (2001). Multielement system design in astronomy and radio science. *Astrophysics and Space Science Library, 268,* 53–70.

Kuhn, H. W., & Tucker, A. (1951). Nonlinear programming. *Proceedings of 2nd Berkeley Symposium* (p. 481). Berkeley: University of California Press.

Labeyrie, A. (1976). High-resolution techniques in optical astronomy. In E. Wolf (Ed.), *Progress in optics* (Vol. XIV, p. 49).

Lantéri, H., & Theys, C. (2005). Restoration of astrophysical images. The case of poisson data with additive Gaussian noise. *EURASIP Journal on Applied Signal Processing, 2500.*

Llacer, J., & Nuñez, J. (1990). In R. L. White & R. J. Allen (Eds.), *The restoration of Hubble space telescope images* (pp. 62–69).

Lucy, L. B. (1974). An iterative technique for the rectification of observed distributions. *The Astronomical Journal, 79,* 745.

Lucy, L. B. (1994). Optimum strategy for inverse problems in statistical astronomy. *Astronomy and Astrophysics, 289,* 983.

Mary, D., Aime, C., & Carlotti, A. (2013). Optimum strategy for inverse problems in statistical astronomy. *EAS Publications Series, 59,* 213.

Nuñez, J., & Llacer, J. (1993). A general bayesian image reconstruction algorithm with entropy prior. Preliminary application to hst data. *Publication of the Astronomical Society of the Pacific, 105,* 1192–1208.

Nuñez, J., & Llacer, J. (1998). Bayesian image reconstruction with space variant noise suppression. *Astronomic and Astrophysics Supplement Series, 131,* 167–180.

Richardson, W. H. (1972). Bayesian-based iterative method of image restoration. *Journal of the Optical Society of America, 62,* 55.

Roddier, F. (1988). Interferometric imaging in optical astronomy. *Physics Report, 170,* 97.

Snyder, D. L., Hammoud, A. M., & White, R. L. (1993). Image recovery from data acquired with charge-coupled-device camera. *Journal of the Optical Society of America, 10,* 1014.

Snyder, D. L., Helstrom, C. W., Lanterman, A. D., Faisal, M., & White, R. L. (1995). Compensation for readout noise in ccd images. *Journal of the Optical Society of America, 12,* 272.

Tikhonov, A., & Arsenin, V. (1976). Fundamentals of statistical signal processing: Detection theory. Moscow: Mir.

Titterington, D. M. (1985). General structure of regularization procedures in image reconstruction. *Astronomy and Astrophysics, 144,* 381–387.

Weigelt, G. (1991). Triple-correlation imaging in optical astronomy. In E. Wolf (Ed.), *Progress in optics* (Vol. XXIX, p. 295).

Wu, C. H., & Anderson, J. M. M. (1997). Novel deblurring algorithms for images captured with CCD cameras. *Journal of the Optical Society of America, 7,* 1421.

Chapter 2
Reconstruction of Solar Subsurfaces by Local Helioseismology

Alexander G. Kosovichev and Junwei Zhao

Abstract Local helioseismology has opened new frontiers in our quest for under-standing of the internal dynamics and dynamo on the Sun. Local helioseismology reconstructs subsurface structures and flows by extracting coherent signals of acoustic waves traveling through the interior and carrying information about subsurface perturbations and flows, from stochastic oscillations observed on the surface. The initial analysis of the subsurface flow maps reconstructed from the 5 years of SDO/HMI data by time-distance helioseismology reveals the great potential for studying and understanding of the dynamics of the quiet Sun and active regions, and the evolution with the solar cycle. In particular, our results show that the emergence and evolution of active regions are accompanied by multi-scale flow patterns, and that the meridional flows display the North-South asymmetry closely correlating with the magnetic activity. The latitudinal variations of the meridional circulation speed, which are probably related to the large-scale converging flows, are mostly confined in shallow subsurface layers. Therefore, these variations do not necessarily affect the magnetic flux transport. The North-South asymmetry is also pronounced in the variations of the differential rotation ('torsional oscillations'). The calculations of a proxy of the subsurface kinetic helicity density show that the helicity does not vary during the solar cycle, and that supergranulation is a likely source of the near-surface helicity.

2.1 Introduction

Observations of solar oscillations provide a unique opportunity to obtain information about the structure and dynamics of the solar interior beneath the visible surface. The oscillations with a characteristic period of 5 min represent acoustic waves stochastically excited by the turbulent convection in a shallow subsurface layer. The excitation mechanism has not been completely understood. However,

A.G. Kosovichev (✉) • J. Zhao
New Jersey Institute of Technology, Newark, NJ 07103, USA

Stanford University, Stanford, CA 95305, USA
e-mail: sasha@bbso.njit.edu

© Springer International Publishing Switzerland 2016
J.-P. Rozelot, C. Neiner (eds.), *Cartography of the Sun and the Stars*,
Lecture Notes in Physics 914, DOI 10.1007/978-3-319-24151-7_2

recent numerical simulations have shown that the waves can be excited due to the interaction of turbulent vortex tubes ubiquitously generated in the intergranular lanes (Kitiashvili et al. 2011). These stochastic waves produce chaotic oscillation patterns on the solar surface. However, a spectral analysis of the time series of these patterns reveals that most of the oscillation power is concentrated in a set of normal modes (Fig. 2.1a). These modes represent standing acoustic waves trapped in the subsurface layers by their reflection between the sharp density gradient near the surface, and the increasing sound speed in the interior. The depth of the inner reflection depends on the horizontal wavelength of the oscillations. The horizontal wavelength, λ_h, is usually represented in terms of the spherical harmonic degree, $\ell = 2\pi R/\lambda_h$. The oscillation frequency is expressed in terms of cyclic frequency $\nu = \omega/2\pi$. In the $\ell - \nu$ diagram shown in Fig. 2.1a, the lowest ridge represents the surface gravity mode (f-mode). The other ridges are acoustic modes of various radial order n, which is equal to the number of nodes along the radius. This number is higher for higher frequency ridges. The time-series of solar oscillations have been obtained almost uninterruptedly since 1995 from the ground-based network GONG and space mission SOHO (Solar and Heliospheric Observatory) and SDO (Solar Dynamics Observatory). The oscillation frequencies are routinely measured from 72- and 108-day time series by fitting the modal lines which are used for

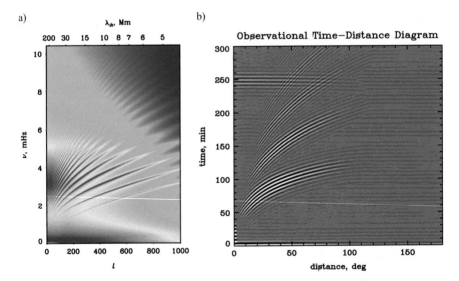

Fig. 2.1 (a) The power spectrum of solar oscillations as a function of the angular degree ℓ, and cyclic frequency, ν. The enhanced power corresponds to the normal oscillation modes of the Sun. (b) The cross-covariance function ('time-distance diagram') of solar oscillations as a function of the distance between the correlation points on the solar surface and the time lag of the cross-covariance. The lowest ridge is formed by acoustic wave packets traveling between two surface points ('source' and 'receiver') through the solar interior (so-called, the first skip); the higher ridges are formed by the wave packets arriving to the receiver after additional reflections from the surface (the 'second' skip, and so on)

inferring variations of the sound speed, asphericity, and differential rotation rate. This approach called 'global helioseismology' has provided important information about the structure, composition and dynamics of the solar interior. In particular, it was led to the discovery of a sharp radial gradient of the differential rotation at the base of the convection zone (Kosovichev 1996b), the so-called tachocline, the near-surface rotational shear layer (Schou et al. 1998), subsurface zonal flows migrating with the solar activity cycle (Kosovichev and Schou 1997). Recent analysis of the high-degree oscillation modes revealed a sharp gradient of the sound speed in a narrow 30 Mm deep layer just beneath the solar surface (Reiter et al. 2015). This layer (called 'leptocline', Godier and Rozelot 2001) presumably plays an important role in the solar dynamo (Pipin and Kosovichev 2011).

It is important to note that while the oscillation power spectrum extends into the high-frequency region (10 mHz and higher), only the ridge parts with the frequency below the acoustic cut-off frequency (which is approximately at 5.2 mHz) represent the normal modes. The higher frequency parts correspond to so-called 'pseudo-modes' . The pseudo-modes are formed by interference between the waves traveling from the excitation sources directly to the surface and the waves which come to the same surface location after reflection in the interior. The pseudo-mode ridges are close to the mode ridges (so that the ridges look continuous) because the excitation sources are located very close to the surface where the oscillations are observed. The pseudo-mode frequencies depend on details of the excitation mechanism and on the wave interaction with the solar atmosphere. Therefore, so-far, only the normal modes have been used for the reconstruction of solar subsurfaces. The primary restriction of global helioseismology is that it can only reconstruct the azimuthally averaged properties of the interior. This is not sufficient for the understanding of the solar dynamics and magnetism.

The three-dimensional structure of the solar subsurfaces can be reconstructed by techniques of local helioseismology. One of these techniques, called 'ring-diagram analysis' (Gough and Toomre 1983) is based on measuring frequency shifts in local (typically 15×15 deg) areas, and uses the global helioseismology description of the mode frequency sensitivity to local sound-speed variations and flows. This techniques allows us to reconstruct the solar subsurfaces with relatively low spatial resolution in shallow regions. The reconstruction with higher spatial resolution and much deeper in the interior can be achieved by methods based on extracting coherent wave signals and measuring variations of the wave travel times or phase shift. These techniques called time-distance helioseismology (Duvall et al. 1993) and acoustic holography (Lindsey and Braun 2000) employ cross-covariance functions of solar oscillations instead of the power spectral analysis. The discovery that coherent signals, such as wave packets, can be extracted from the cross-covariance functions of the stochastic solar oscillations (Fig. 2.1b) was made by Duvall et al. (1993). This approach was then developed in helioseismology, terrestrial seismology, and other disciplines, and in broader applications is called 'ambient noise imaging'. The foundation of this approach is based on the property of cross-covariance functions to represent wave signals corresponding to point sources. Roughly speaking the cross-covariance function can be considered as the Green's function of the solar wave equation. In real solar conditions this is only an approximation because of the limited frequency bandwidth of solar oscillations and inhomogeneities of the solar structures and distribution of the stochastic sources. A complete theory of this approach of helioseismology has not been developed. It requires extensive studies of wave interaction with turbulence, flows and magnetic field. Nevertheless, the initial results based on relatively simple descriptions of wave propagation have provided important insights into the three-dimensional structures and flow patterns of the solar subsurfaces. The primary focus of these studies is mapping the flow patterns associated with the solar cycle, and formation and evolution of active regions.

In the current state of local helioseismology the systematic errors as well as effects of the stochastic realization noise have not been fully investigated. These studies require substantial effort for modeling the wave dynamics in realistic solar conditions, and require 3D MHD simulations on large supercomputer systems. The validation and testing of the time-distance technique have been performed by comparing the helioseismic inversions in the shallowest layer with the surface flows obtained by a local correlation tracking technique (Liu et al. 2013), and through the analysis and inversion of numerical simulation data for subsurface flows and sound-speed variations (Birch et al. 2011; Hartlep et al. 2013; Parchevsky and Kosovichev 2009; Parchevsky et al. 2014). The testing for regions with strong magnetic field has not yet been completed. However, the simulations of the wave propagation in sunspot models showed that one of the primary effects in sunspot regions is the wave reflection from deeper layers, compared to the quiet-Sun regions, where the plasma parameter, $\beta = 8\pi P/B^2$, the ratio of the gas pressure to magnetic pressure, is equal to unity (this layer also corresponds to the deeper photospheric surface of sunspots, known as the Wilson depression). Below the Wilson depression level the gas pressure dominates, and the helioseismic acoustic waves behave like fast MHD waves: the wave speed becomes anisotropic, and also depends on the temperature stratification beneath sunspots. The magnetic and temperature effects have not been separated in the wave-speed inversion results (Kosovichev et al. 2000). This is an important task of local helioseismology. One of the difficulties is that the limited computer power has not allowed to simulated the sunspot models sufficiently large and deep for the helioseismology testing, so that the wave properties are not affected by the boundary conditions of the simulations. A comparison of the wave-speed inversions obtained for a sunspot regions by the time-distance and ring-diagram techniques (see Kosovichev 2012; Kosovichev et al. 2011 and references therein) shows a good qualitative agreement: both inversions show a two-layer structure with a layer of reduced wave speed beneath the surface followed by a layer of an increased wave speed. However, the depth of these layers is different, perhaps, because of the drastically different spatial resolution, and different contributions of magnetic field. A comparison of the time-distance and acoustic holography inversion results has been performed by using artificial simulation data (Birch et al. 2011; Parchevsky et al. 2014).

2.2 Time-Distance Helioseismology from SDO

This brief review presents recent results obtained by the helioseismology reconstructions of subsurface flows in the near-surface layer and development of active
regions. The results are obtained by analyzing inversions for subsurface flows from
the SDO Joint Science Operations Center (JSOC) at Stanford University (Scherrer
et al. 2012). The JSOC data analysis pipeline provides 3D maps of solar flows
covering almost the whole disk (within 60 deg from the disk center) in the range
of depths from 0 to 30 Mm, every 8 h.

The time-distance helioseismology pipeline (Fig. 2.2a) developed by the Stanford
group (Couvidat et al. 2012; Zhao et al. 2012) utilizes two different methods for

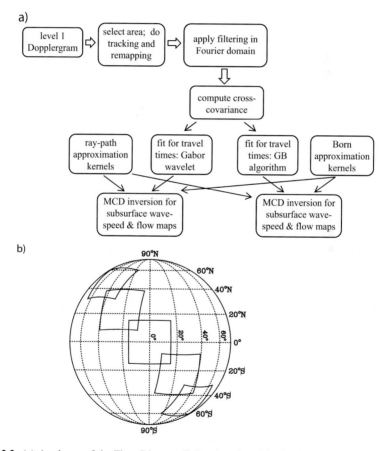

Fig. 2.2 (**a**) A scheme of the Time-Distance Helioseismology Pipeline implemented at the Joint
Science Operations Center (JSOC) for Solar Dynamics Observatory at Stanford University (Zhao
et al. 2012); (**b**) Illustration of the surface locations of the individual patches used for inferences of
the subsurface structure and flows; the total 25 patches are used to cover 120 × 120 deg of the disk
area

measuring the acoustic travel times: (1) the method of fitting the Gabor wavelet to the cross-covariance function, which provides measurements of both the phase and group travel times (Kosovichev and Duvall 1997) and (2) the method of calculating the travel-time shift relative to a reference cross-covariance function (Gizon and Birch 2002), usually calculated for a quiet-Sun region. The two sets of the travel times are calculated independently for 11 travel distances, for the same 25 areas covering the solar disk (Fig. 2.2b), and for the same 8-h intervals. Then, the travel times are used for reconstruction of subsurface flows in 11 subsurface layers in the depth ranges: 1–3, 3–5, 5–7, 7–10, 10–13, 13–17, 17–21, 21–26, 26–30, and 30–35 Mm, and with the horizontal spatial sampling of 0.12 deg (1.5 Mm).

The inversions are performed by using two different methods for calculating the travel-time sensitivity functions: (1) the ray-path approximation (Kosovichev 1996a; Kosovichev and Duvall 1997) and (2) the first Born approximation (Birch and Kosovichev 2000, 2001; Birch et al. 2001, 2004; Birch and Gizon 2007). The inversions are performed using the Multi-Channel Deconvolution (MCD) technique (Jacobsen et al. 1999) for the two independent travel-time measurements using the two types of the sensitivity kernels. Therefore, the pipeline output consists of four sets of subsurface flow maps for the same areas on the Sun (Zhao et al. 2012). This allows the comparison of the different approaches and estimate potential systematic errors. The reconstruction of subsurface flows has also been tested through analysis and inversion of numerical simulation results as well as by the comparison of the flow maps obtained by the different techniques, and also by comparing the inversion results in the shallowest layer with the surface flows measured by the feature correlation tracking techniques (Liu et al. 2013). Figure 2.3 illustrates the comparison of the flow maps below a sunspot region, obtained by using two different techniques for measuring the travel times and two different models for the travel-time sensitivity functions. The results show that the agreement is quite good everywhere except the areas close to the sunspot. As discussed in the Introduction, the effects of a strong magnetic field and large perturbations of the thermodynamic structure have not been fully investigated. Solving this problem requires more studies of systematic uncertainties using realistic numerical simulations. Nevertheless, the currently available inferences shed light on the intriguing dynamics of the solar interior.

2.3   Subsurface Flows and Effects of Solar Activity

An example of the reconstructed subsurface flow maps is illustrated in Fig. 2.4a, which shows the distribution of the divergence of the horizontal flow velocity in the depth range 1–3 Mm. The primary feature covering the whole surface is supergranulation. The outflows of a few hundred m/s in the supergranulation cells are represented by light dot-like features. However, the examination of these maps shows that in the vicinity of magnetic active regions (shown in Fig. 2.4b), the supergranulation pattern is substantially suppressed. The flow pattern beneath and

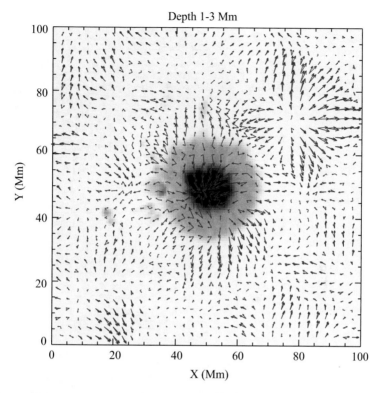

Fig. 2.3 Comparison of the subsurface flow maps in the depth range from 1 to 3 Mm, obtained by using two different types of the travel measurements and two different approaches for calculation of the travel-time sensitivity functions: *red arrows* show the flow field obtained by using the Gabor-wavelet fitting technique and the raypath kernels (Kosovichev and Duvall 1997), the *blue arrows* are obtained by using the cross-correlation approach for the travel times (Gizon and Birch 2002), and the Born-approximation kernels (Birch and Kosovichev 2000). The *longest arrows* correspond to the velocity of about 1 km/s

around sunspots represents a complicated combination of converging flows towards the sunspot centers (displayed as dark dots in the divergence map) surrounded by outflows (represented as white rings). Such a pattern, similar to the flows shown in Figs. 2.3 and 2.5, has been previously studied using SOHO/MDI and Hinode data (Zhao et al. 2001, 2009, 2010). A new feature of the SDO/HMI analysis is that the HMI data allow us to reconstruct the flows in a shallow subsurface layer, and match these to the directly observed surface flows. This agreement provides more confidence in the helioseismic inferences.

Figure 2.5 shows a portion of the horizontal velocity map around an emerging active region NOAA 11726, during its development phase. This is the largest active region observed by the HMI instrument during the first 5 years of operation. The flow velocities are shown by arrows, and the photospheric magnetogram is represented by the color map. Such flow maps are obtained with 1-h sampling,

Fig. 2.4 (**a**) A full-disk map
showing the divergence of the
horizontal velocity at the
depth of 1–3 Mm, obtained on
December 19, 2014,
12:00 UT. The *bright
point-like areas* represent
diverging supergranulation
flows, the *dark areas*
surrounded by *bright rings*
represent flows converging
beneath sunspots and
diverging in the areas
surrounding the sunspots. (**b**)
The corresponding maps of
the line-of-sight magnetic
field obtained from the SDO
Helioseismic and Magnetic
Imager

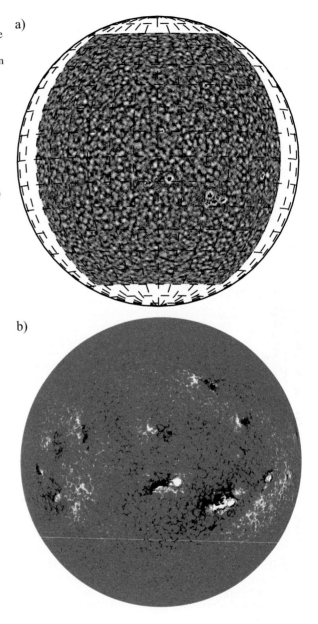

although each map requires 8-h time series of Dopplergrams for the helioseismology
analysis. The analysis of these maps indicates that the converging flows beneath
the sunspots are developed simultaneously with the sunspot formation, as was
previously found from analysis of the SOHO/MDI data (Kosovichev and Duvall
2006; Kosovichev 2009), and, probably, is closely associated with the mechanism of
the sunspot formation. At the same time, large-scale diverging flows are developed

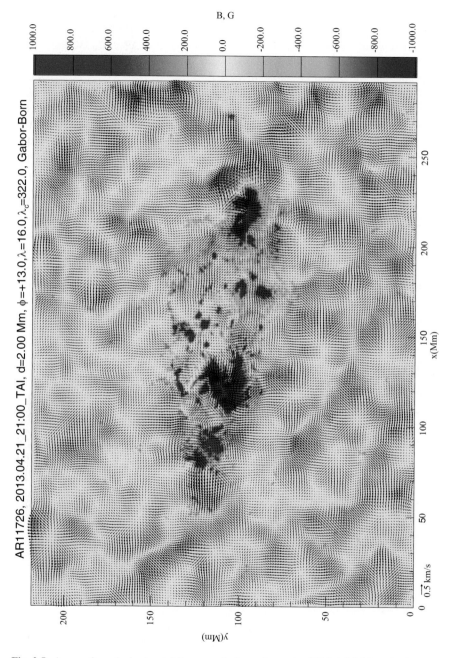

Fig. 2.5 *Arrows* show the horizontal flow map in the active region NOAA 11726 at the depth of 2 Mm on April 21, 2013, 21:00 UT, about 2 days after its emergence from the interior. The *color background* image shows the surface magnetic field. The flow map is reconstructed by using the Gabor-wavelet technique for measuring the travel times, and the Born-approximation sensitivity kernels for the flow velocity inversion

around the active region, and are probably related to the well-known phenomenon of the surface 'moat' flow.

Outside the active region the flow pattern is mostly represented by supergranulation which, however, is clearly disturbed by the presence of the active region. It is interesting that the spatial averaging of these flow maps reveals a large-scale pattern of converging flows occupying a surrounding area which is significantly larger than the active region (Fig. 2.6). Such converging flows with the characteristic speed of about 50 m/s were first discovered by the ring-diagram technique (Haber et al. 2003). The origin of these flows is not understood, but our analysis shows that these flows are formed and stable only when the active region is fully developed, and, thus, they are not associated with the emergence of magnetic flux and formation of the active region.

The large-scale converging flows around active regions play an important role in the solar-cycle evolution of the meridional circulation (Haber et al. 2002; Zhao and Kosovichev 2004). The meridional circulation can be calculated from the reconstructed subsurface flow maps by averaging the North-South component of the flow velocity. Figure 2.7a shows a map of the North-South velocity component smoothed with a 5-deg Gaussian window.

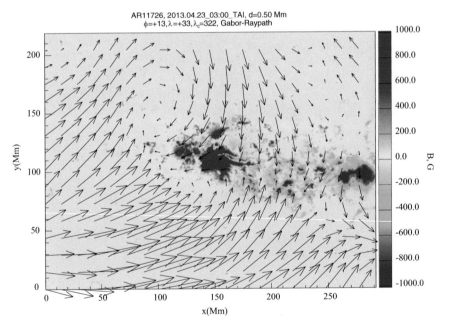

Fig. 2.6 *Arrows* show the horizontal flow map around the active region NOAA 11726 at the depth of 0.3 Mm on April 23, 2013, 03:00 UT (when the active region is fully developed) after averaging the high-resolution flow map on a grid with a 15-degree sampling. The *color background image* shows the surface magnetic field. The typical flow speed is about 50 m/s

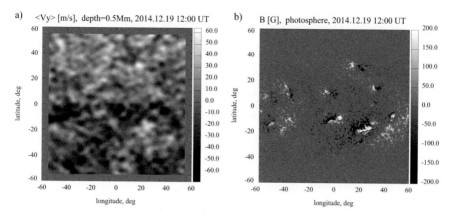

Fig. 2.7 (**a**) A map of the North-South component of the subsurface flow velocity of December 12, 2014, 12:00 UT, smoothed with a 5 deg Gaussian window, reveals that the meridional flow pattern disturbed the converging flows around active regions. (**b**) The corresponding photospheric magnetogram map. *Red color* shows the positive polarity, the *blue color* shows the negative polarity

The appearance of the poleward trends in each hemisphere is apparent. It is interesting that the meridional circulation can be detected in a single flow map, but it is also important that the flow pattern correlates with the surface magnetic field map shown in Fig. 2.7b. The variations can be interpreted as caused by the large-scale converging flows around active regions. However, the strongest variations in this flow map seem to be in the areas of decaying active regions. Thus, it is important to investigate the formation and evolution of the converging flows during the whole evolution of active regions, from their formation to decay.

The evolution of meridional circulation is obtained by averaging the North-South velocity component over longitude and 1-month periods, and displaying the averages in the form of a time-latitude diagram (Fig. 2.8a). This diagram shows that the evolution of the subsurface meridional circulation correlates with the magnetic activity in each hemisphere. At the beginning of the current cycle most active regions emerged in the Northern hemisphere, where we see a strong variation of the meridional circulation speed: a sharp increase at low latitudes (in the 10–20 deg interval) and a decrease at mid latitudes (in the 20–30 deg range). A similar variation in the Southern hemisphere is observed in 2014–2015 when most magnetic activity was in the South (Fig. 2.9b).

Such variations of the meridional circulation may affect the magnetic flux transport and the polar magnetic field polarity reversal. However, this link has not been fully established (Švanda et al. 2007a,b, 2008). Figure 2.8b shows the variation of the mean (averaged over the whole period) meridional circulation profile with depth. It appears that at a depth of ∼10 Mm the latitudinal variations are significantly reduced, and at the depth of ∼20 Mm almost entirely disappear. Therefore, if the large-scale magnetic flux is anchored at this depth or lower then its transport is not affected by the meridional flow variations.

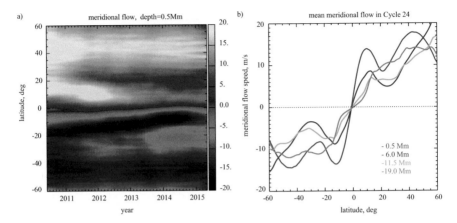

Fig. 2.8 (**a**) Evolution of the subsurface meridional flows obtained from the 5-years of the SDO/HMI observations during Solar Cycle 24. The *red* and *yellow colors* show the flow components towards the North pole, the *green* and *blue colors* show the South-ward flow. The color scale range is from −20 to 20 m/s. (**b**) The mean meridional flow averaged for the whole period of observations at four different depths

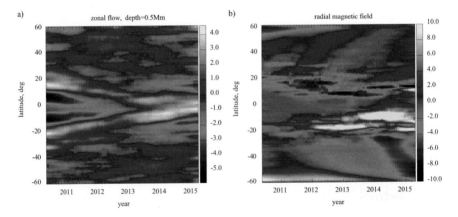

Fig. 2.9 (**a**) The evolution of the subsurface zonal flows ('torsional oscillation') during the first 5 years of the SDO observations, covering the raising phase of Solar Cycle 24. The flow map (in m/s) is obtained by averaging of the azimuthal flow component of longitude and 1-month time, subtracting the mean rotational velocity from each of the 1-month averaged profile, and then stacking the residual velocity profiles and smoothing with 1-year running window to remove the annual variations due to the inclination of the Earth orbit. The *yellow* and *red colors* correspond to the zonal flows faster than the mean solar rotation at the same depth, and the *blue color* shows the slower rotating regions. (**b**) The corresponding magnetic 'butterfly' diagram from the SDO/HMI data, showing the evolution of the mean radial magnetic field (in G) in the solar photosphere during the same period

Solar-cycle variations of the differential rotation are also of great interest for the understanding of the mechanisms of solar activity. These variations, known as 'torsional oscillation', have been detected from the surface Doppler-shift maps (Howard and Labonte 1980; Ulrich 2001), and by global (Kosovichev and Schou 1997) and local (Zhao and Kosovichev 2004) helioseismology. The high-resolution flow maps from SDO/HMI provide new opportunities for investigating the detailed structure and evolution of these flows. Figure 2.9a shows the time-latitude diagram of variations of the differential rotation during the 5 years of the SDO/HMI observations. These variations are relative to the mean differential rotation profile averaged for the whole period, and smoothed with a 1-year window to remove the orbital variations. As it was established before, the zonal flow closely correlates with the magnetic butterfly diagram (Fig. 2.9b). However, our results also show the North-South asymmetry of the flows, which follows the asymmetry of the magnetic activity.

The flow maps allow us to investigate other important properties of the subsurface dynamics of the Sun, which previously were not accessible. For illustration, in Fig. 2.10a we show a map of the kinetic helicity proxy calculated as $\nabla \boldsymbol{v}_h \cdot (\nabla \times \boldsymbol{v}_h)_z$, where $\boldsymbol{v}_h$ is the horizontal velocity component. By looking at this map one can notice that the Northern hemisphere is darker than the Southern hemisphere, and that the asymmetry is particularly pronounced in the supergranulation cells. After the longitudinal and time averaging of the individual helicity proxy maps we obtain the time-latitude diagram (Fig. 2.10b), which shows that the kinetic helicity does not vary on this time scale. This result puts constraints on the dynamo theories, and also shows that the supergranulation flows are likely a primary source of the near-surface helicity.

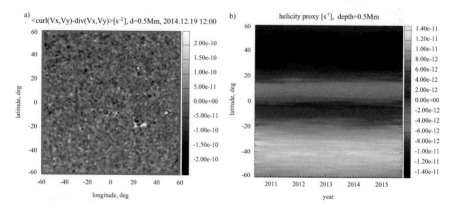

Fig. 2.10 (**a**) The proxy of kinetic helicity density, $\nabla \boldsymbol{v}_h \cdot (\nabla \times \boldsymbol{v}_h)_z$, calculated from the flow map of December 12, 2014, 12:00 UT, reveals a systematic North-South asymmetry in local sources associated with supergranulation. (**b**) The evolution of the mean helicity proxy during the observed period of Solar Cycle 24

2.4   Conclusion

The initial analysis of the subsurface flow maps reconstructed from the 5 years of SDO/HMI data by time-distance helioseismology reveals the great potential for studying and understanding the dynamics of the quiet-Sun and active regions, and the evolution with the solar cycle. In particular, our results show that the emergence and evolution of active regions are accompanied by multi-scale flow patterns. Beneath the sunspot, during their formation, we observe appearance of flows converging towards the sunspot center. and also the 'moat'-like flows diverging from the active region in the surrounding regions. On the larger scale, revealed by averaging the high-resolution flow maps, we find a pattern of flows converging towards the active region. This pattern is formed when the active region is fully developed. On the global-Sun scale, the flow maps allow us to investigate the structure and evolution of the meridional flows. In particular, we find that the meridional flows display the North-South asymmetry closely correlating with the magnetic activity. The latitudinal variations of the meridional circulation speed, which are probably related to the large-scale converging flows, are mostly confined in a shallow subsurface layers. Therefore, these variations do not necessarily affect the magnetic flux transport. The North-South asymmetry is also pronounced in the variations of the differential rotation ('torsional oscillation'). The calculations of a proxy of the subsurface kinetic helicity density show that the helicity does not vary during the solar cycle, and that the supergranulation is a likely source of the near-surface helicity. These initial results are obtained from the analysis of a small sample of flow maps produced by the SDO/HMI time-distance helioseismology pipeline. Further detailed investigations are required for understanding the complicated subsurface dynamics of the Sun.

Acknowledgements This work was supported by the CNRS, and NASA grants NNX09AJ85G and NNX14AB70G.

References

Birch, A. C., & Gizon, L. (2007). Linear sensitivity of helioseismic travel times to local flows. *Astronomische Nachrichten, 328*, 228. doi:10.1002/asna.200610724. 1002.2338.

Birch, A. C., & Kosovichev, A. G. (2000). Travel time sensitivity kernels. *Solar Physics, 192*, 193–201. doi:10.1023/A:1005283526062. http://www.adsabs.harvard.edu/abs/2000SoPh..192..193B.

Birch, A. C., & Kosovichev, A. G. (2001). The born approximation in time-distance helioseismology. In A. Wilson & P. L. Pallé (Eds.), *SOHO 10/GONG 2000 Workshop: Helio- and Asteroseismology at the Dawn of the Millennium* (Vol. 464, pp. 187–190). ESA Special Publication. http://www.adsabs.harvard.edu/abs/2001ESASP.464..187B.

Birch, A. C., Kosovichev, A. G., Price, G. H., & Schlottmann, R. B. (2001). The accuracy of the born and ray approximations in time-distance helioseismology. *The Astrophysical Journal, 561*, L229–L232. doi:10.1086/324766. http://www.adsabs.harvard.edu/abs/2001ApJ...561L.229B.

Birch, A. C., Kosovichev, A. G., & Duvall, T. L., Jr., (2004). Sensitivity of acoustic wave travel times to sound-speed perturbations in the solar interior. *The Astrophysical Journal, 608*, 580–600. doi:10.1086/386361. http://www.adsabs.harvard.edu/abs/2004ApJ...608..580B.

Birch, A. C., Parchevsky, K. V., Braun, D. C., & Kosovichev, A. G. (2011). "Hare and hounds" tests of helioseismic holography. *Solar Physics, 272*, 11–28. doi:10.1007/s11207-011-9799-1. http://www.adsabs.harvard.edu/abs/2011SoPh..272...11B.

Couvidat, S., Zhao, J., Birch, A. C., Kosovichev, A. G., Duvall, T. L., Parchevsky, K., et al. (2012). Implementation and comparison of acoustic travel-time measurement procedures for the Solar Dynamics Observatory/Helioseismic and Magnetic Imager time – distance helioseismology pipeline. *Solar Physics, 275*, 357–374. doi:10.1007/s11207-010-9652-y. http://www.adsabs.harvard.edu/abs/2012SoPh..275..357C.

Duvall, T. L., Jr., Jefferies, S. M., Harvey, J. W., & Pomerantz, M. A. (1993). Time-distance helioseismology. *Nature, 362*, 430–432. doi:10.1038/362430a0.

Gizon, L., & Birch, A. C. (2002). Time-distance helioseismology: The forward problem for random distributed sources. *The Astrophysical Journal, 571*, 966–986. doi:10.1086/340015.

Godier, S., & Rozelot, J. P. (2001). A new outlook on the 'differential theory' of the solar quadrupole moment and oblateness. *Solar Physics, 199*, 217–229. doi:10.1023/A: 1010354901960. http://www.adsabs.harvard.edu/abs/2001SoPh..199..217G.

Gough, D. O., & Toomre, J. (1983). On the detection of subphotospheric convective velocities and temperature fluctuations. *Solar Physics, 82*, 401–410. doi:10.1007/BF00145579.

Haber, D. A., Hindman, B. W., Toomre, J., Bogart, R. S., Larsen, R. M., & Hill, F. (2002). Evolving submerged meridional circulation cells within the upper convection zone revealed by ring-diagram analysis. *The Astrophysical Journal, 570*, 855–864. doi:10.1086/339631.

Haber, D. A., Hindman, B. W., & Toomre, J. (2003). Interaction of solar subsurface flows with major active regions. In H. Sawaya-Lacoste (Ed.), *GONG+ 2002. Local and global helioseismology: The present and future* (Vol. 517, pp. 103–108). ESA Special Publication

Hartlep, T., Zhao, J., Kosovichev, A. G., & Mansour, N. N. (2013). Solar wave-field simulation for testing prospects of helioseismic measurements of deep meridional flows. *The Astrophysical Journal, 762*, 132. doi:10.1088/0004-637X/762/2/132. http://www.adsabs.harvard.edu/abs/2013ApJ...762..132H. 1209.4602.

Howard, R., & Labonte, B. J. (1980). The Sun is observed to be a torsional oscillator with a period of 11 years. *The Astrophysical Journal, 239*, L33–L36. doi:10.1086/183286.

Jacobsen, B., Moller, I., Jensen, J., & Efferso, F. (1999). Multichannel deconvolution, mcd, in geophysics and helioseismology. *Physics and Chemistry of the Earth A, 24*, 215–220. doi: 10.1016/S1464-1895(99)00021-6.

Kitiashvili, I. N., Kosovichev, A. G., Mansour, N. N., & Wray, A. A. (2011). Excitation of acoustic waves by vortices in the quiet Sun. *The Astrophysical Journal, 727*, L50. doi:10.1088/ 2041-8205/727/2/L50. http://www.adsabs.harvard.edu/abs/2011ApJ...727L..50K. 1011.3775.

Kosovichev, A. G. (1996a). Tomographic imaging of the Sun's interior. *The Astrophysical Journal, 461*, L55. doi:10.1086/309989. http://www.adsabs.harvard.edu/abs/1996ApJ...461L..55K.

Kosovichev, A. G. (1996b). Helioseismic constraints on the gradient of angular velocity at the base of the solar convection zone. *The Astrophysical Journal, 469*, L61. doi:10.1086/310253. http:// www.adsabs.harvard.edu/abs/1996ApJ...469L..61K.

Kosovichev, A. G. (2009). Photospheric and subphotospheric dynamics of emerging magnetic flux. *Space Science Reviews, 144*, 175–195. doi:10.1007/s11214-009-9487-8. http://www.adsabs. harvard.edu/abs/2009SSRv..144..175K. 0901.0035.

Kosovichev, A. G. (2012). Local helioseismology of sunspots: Current status and perspectives. *Solar Physics, 279*, 323–348. doi:10.1007/s11207-012-9996-6. http://www.adsabs.harvard. edu/abs/2012SoPh..279..323K.

Kosovichev, A. G., & Duvall, T. L. (2006). Active region dynamics. *Space Science Reviews, 124*, 1–12. doi:10.1007/s11214-006-9112-z. http://www.adsabs.harvard.edu/abs/2006SSRv..124.... 1K.

Kosovichev, A. G., & Duvall, T. L., Jr., (1997). Acoustic tomography of solar convective flows and structures. In F. P. Pijpers, J. Christensen-Dalsgaard, & C. S. Rosenthal (Eds.),

SCORe'96: Solar Convection and Oscillations and Their Relationship (Vol. 225, pp. 241–260). Astrophysics and Space Science Library. http://www.adsabs.harvard.edu/abs/1997ASSL..225.. 241K.

Kosovichev, A. G., & Schou, J. (1997). Detection of zonal shear flows beneath the Sun's surface from f-mode frequency splitting. *The Astrophysical Journal, 482*, L207–L210. doi:10.1086/ 310708. http://www.adsabs.harvard.edu/abs/1997ApJ...482L.207K.

Kosovichev, A. G., Duvall, T. L., Jr., & Scherrer, P. H. (2000). Time-distance inversion methods and results – (invited review). *Solar Physics, 192*, 159–176. doi:10.1023/A:1005251208431. http://www.adsabs.harvard.edu/abs/2000SoPh..192..159K.

Kosovichev, A. G., Basu, S., Bogart, R., Duvall, T. L., Jr., Gonzalez-Hernandez, I., Haber, D., et al. (2011). Local helioseismology of sunspot regions: Comparison of ring-diagram and time-distance results. *Journal of Physics Conference Series, 271*(1), 012005. doi:10.1088/ 1742-6596/271/1/012005. http://www.adsabs.harvard.edu/abs/2011JPhCS.271a2005K. 1011. 0799.

Lindsey, C., & Braun, D. C. (2000). Basic principles of solar acoustic holography – (invited review). *Solar Physics, 192*, 261–284. doi:10.1023/A:1005227200911.

Liu, Y., Zhao, J., & Schuck, P. W. (2013). Horizontal flows in the photosphere and subphotosphere of two active regions. *Solar Physics, 287*, 279–291. doi:10.1007/s11207-012-0089-3.

Parchevsky, K. V., & Kosovichev, A. G. (2009). Numerical simulation of excitation and propagation of helioseismic MHD waves: Effects of inclined magnetic field. *The Astrophysical Journal, 694*, 573–581. doi:10.1088/0004-637X/694/1/573. http://www.adsabs.harvard.edu/ abs/2009ApJ...694..573P. 0806.2897.

Parchevsky, K. V., Zhao, J., Hartlep, T., & Kosovichev, A. G. (2014). Verification of the helioseismology travel-time measurement technique and the inversion procedure for sound speed using artificial data. *The Astrophysical Journal, 785*, 40. doi:10.1088/0004-637X/785/1/40. http:// www.adsabs.harvard.edu/abs/2014ApJ...785...40P.

Pipin, V. V., & Kosovichev, A. G. (2011). The subsurface-shear-shaped solar $\alpha\Omega$ dynamo. *The Astrophysical Journal, 727*, L45. doi:10.1088/2041-8205/727/2/L45. http://www.adsabs. harvard.edu/abs/2011ApJ...727L..45P. 1011.4276.

Reiter, J., Rhodes, E. J., Jr., Kosovichev, A. G., Schou, J., Scherrer, P. H., & Larson, T. P. (2015). A method for the estimation of p-mode parameters from averaged solar oscillation power spectra. *The Astrophysical Journal, 803*, 92. doi:10.1088/0004-637X/803/2/92. http://www. adsabs.harvard.edu/abs/2015ApJ...803...92R. 1504.07493.

Scherrer, P. H., Schou, J., Bush, R. I., Kosovichev, A. G., Bogart, R. S., Hoeksema, J. T., et al. (2012). The Helioseismic and Magnetic Imager (HMI) investigation for the Solar Dynamics Observatory (SDO). *Solar Physics, 275*, 207–227. doi:10.1007/s11207-011-9834-2. http:// www.adsabs.harvard.edu/abs/2012SoPh..275..207S.

Schou, J., Antia, H. M., Basu, S., Bogart, R. S., Bush, R. I., Chitre, S. M., et al. (1998) Helioseismic studies of differential rotation in the solar envelope by the solar oscillations investigation using the Michelson Doppler Imager. *The Astrophysical Journal, 505*, 390–417. doi:10.1086/306146. http://www.adsabs.harvard.edu/abs/1998ApJ...505..390S.

Ulrich, R. K. (2001). Very long lived wave patterns detected in the solar surface velocity signal. *The Astrophysical Journal, 560*, 466–475. doi:10.1086/322524.

Švanda, M., Kosovichev, A. G., & Zhao, J. (2007a). Speed of meridional flows and magnetic flux transport on the Sun. *The Astrophysical Journal, 670*, L69–L72. doi:10.1086/524059. http:// www.adsabs.harvard.edu/abs/2007ApJ...670L..69S. 0710.0590.

Švanda, M., Zhao, J., & Kosovichev, A. G. (2007b) Comparison of large-scale flows on the Sun measured by time-distance helioseismology and local correlation tracking. *Solar Physics, 241*, 27–37. doi:10.1007/s11207-007-0333-4. http://www.adsabs.harvard.edu/abs/2007SoPh..241... 27S. astro-ph/0701717.

Švanda, M., Kosovichev, A. G., & Zhao, J. (2008). Effects of solar active regions on meridional flows. *The Astrophysical Journal, 680*, L161–L164. doi:10.1086/589997. http://www.adsabs. harvard.edu/abs/2008ApJ...680L.161S. 0805.1789.

Zhao, J., & Kosovichev, A. G. (2004). Torsional oscillation, meridional flows, and vorticity inferred in the upper convection zone of the Sun by time-distance helioseismology. *The Astrophysical Journal, 603*, 776–784. doi:10.1086/381489. http://www.adsabs.harvard.edu/abs/2004ApJ...603..776Z.

Zhao, J., Kosovichev, A. G., & Duvall T. L., Jr. (2001). Investigation of mass flows beneath a sunspot by time-distance helioseismology. *The Astrophysical Journal, 557*, 384–388. doi:10.1086/321491. http://www.adsabs.harvard.edu/abs/2001ApJ...557..384Z.

Zhao, J., Kosovichev, A. G., & Sekii, T. (2009). Subsurface structures and flow fields of an active region observed by Hinode. In B. Lites, M. Cheung, T. Magara, J. Mariska, & K. Reeves (Eds.), *The Second Hinode Science Meeting: Beyond Discovery-Toward Understanding.* Astronomical Society of the Pacific Conference Series (Vol. 415, p. 411). http://www.adsabs.harvard.edu/abs/2009ASPC..415..411Z.

Zhao, J., Kosovichev, A. G., & Sekii, T. (2010). High-resolution helioseismic imaging of subsurface structures and flows of a solar active region observed by Hinode. *The Astrophysical Journal, 708*, 304–313. doi:10.1088/0004-637X/708/1/304. http://www.adsabs.harvard.edu/abs/2010ApJ...708..304Z. 0911.1161.

Zhao, J., Couvidat, S., Bogart, R. S., Parchevsky, K. V., Birch, A. C., Duvall, T. L., et al. (2012) Time-distance helioseismology data-analysis pipeline for Helioseismic and Magnetic Imager onboard Solar Dynamics Observatory (SDO/HMI) and its initial results. *Solar Physics, 275*, 375–390. doi:10.1007/s11207-011-9757-y. http://www.adsabs.harvard.edu/abs/2012SoPh..275..375Z. 1103.4646.

Chapter 3
Imaging Surface Spots from Space-Borne Photometry

A.F. Lanza

Abstract A general introduction to the foundations of spot modelling is given. It considers geometric models of the surface brightness distribution in late-type stars as can be derived from their wide-band optical light curves. Spot modelling is becoming more and more important thanks to the high-precision, high duty-cycle photometric time series made available by space-borne telescopes designed to search for planets through the method of transits. I review approaches based on a few spots as well as more sophisticated techniques that assume a continuous distributions of active regions and adopt regularization methods developed to solve ill-posed problems. The use of transit light curves to map spots occulted by a planet as it moves across the disc of its host star is also briefly described. In all the cases, the main emphasis is on the basic principles of the modelling techniques and on their testing rather than on the results obtained from their application.

3.1 Introduction

The photosphere of the Sun is not homogeneous. Dark features, called sunspots, appear and evolve during most of the time, while bright faculae are often observed in proximity to sunspots when they are close to the limb. Those structures are due to the interaction of convection with localized magnetic fields. The total sunspot area does not exceed 0.2–0.3 % of the solar surface. The total facular area can be about one order of magnitude larger, but faculae have a very low contrast close to the disc centre and may not be easily detected there (e.g., Chapman et al. 2001, 2011). Looking at the photosphere with a resolution of the order of $10^2 - 10^3$ km, we see other brightness inhomogeneities associated with magnetic flux tubes that are localized around the borders of the convective cells. In particular, the flux tubes observed around supergranules are brighter than the unperturbed photosphere and form the photospheric network that is best detected on high-resolution magnetograms.

A.F. Lanza (✉)
INAF-Osservatorio Astrofisico di Catania, Via S. Sofia 78, 95123 Catania, Italy
e-mail: nuccio.lanza@oact.inaf.it

© Springer International Publishing Switzerland 2016

J.-P. Rozelot, C. Neiner (eds.), *Cartography of the Sun and the Stars*,
Lecture Notes in Physics 914, DOI 10.1007/978-3-319-24151-7_3

43

In distant late-type stars, we observe similar photospheric features because those stars have surface convection and magnetic fields produced by a large-scale dynamo, at least if they rotate sufficiently fast (e.g., Berdyugina 2005; Strassmeier 2009; Kővári and Oláh 2014). However, the lack of spatial resolution means that we can detect them only indirectly. During this school several methods to reach this goal have been introduced. Here I focus on the information that can be extracted from wide-band photometry, especially from the large datasets recently made available by space-borne telescopes designed to look for planetary transits such as CoRoT and Kepler (e.g., Auvergne et al. 2009; Borucki et al. 2010). The typical accuracy of the measured flux is of the order of 20 parts per million (ppm) on a $V = 12$ magnitude G-type star in 1 h of integration time, considering that Kepler has a telescope diameter of 95 cm.

The rotation of a star changes the projected area of its surface brightness inhomogeneities leading to a rotational modulation of its optical flux. Moreover, the intrinsic evolution of these inhomogeneities contributes to the flux variations. In principle, it is possible to measure the rotation period from the light modulation provided that the inhomogeneities evolve on a timescale long in comparison with the rotation period. If they evolve on a shorter timescale, the light variations will provide information on their typical lifetime, but they cannot be used as tracers to measure the stellar rotation period. This is the case of the Sun. The time variation of its total irradiance is dominated by active regions produced by magnetic fields. When the modulation is dominated by faculae with a typical lifetime of 50–80 days, i.e., 2–3 rotations, we can apply time-series analysis techniques to measure the rotation period. On the other hand, when sunspots with a lifetime of only 10–15 days dominate, the measurement of the rotation period becomes difficult and imprecise (cf. Lanza et al. 2004).

Several techniques were introduced to analyze time series of stellar optical photometry to derive the rotation period, the longitudes where surface inhomo-geneities preferentially form, and their evolution timescales as well as the long-term variations associated with stellar activity cycles, i.e., the phenomena analogous to the 11-year sunspot cycle (e.g., Jetsu 1996; Donahue et al. 1997a,b; Kolláth and Oláh 2009; Lehtinen et al. 2011; Lindborg et al. 2013; McQuillan et al. 2013; Reinhold et al. 2013; McQuillan et al. 2014). We shall not discuss those techniques here, but shall focus on the foundations of the modelling of the rotational modulation by means of a simple geometrical approach (*spot modelling*). In the case of close eclipsing binaries, we can also exploit the occultation of one component star by the other to scan its disc (*eclipse mapping*, e.g., Collier Cameron 1997; Lanza et al. 1998). This approach becomes particularly powerful when a planet transits across the disc of its parent star. Thanks to its small size in comparison to the stellar disc, a detailed scanning of the occulted band becomes possible (*transit mapping*, e.g., Schneider 2000; Silva 2003). The foundations of such a technique will be also briefly reviewed.

3.2 Spot Modelling with Discrete Spots

3.2.1 Model Geometry

For the sake of simplicity, let us consider a single, spherical star rotating with a uniform angular velocity Ω. We assume a Cartesian reference frame fixed in the inertial space (i.e., a non-rotating frame) with the origin O at the barycentre of the star, the z axis along the stellar spin axis ($\hat{z} \equiv \hat{\Omega}$), and the x and y axes in the equatorial plane; the x axis is chosen so that the direction pointing towards the observer $\hat{s} \equiv \widehat{OE}$ is contained in the xz plane (see Fig. 3.1).

Let us consider a point $P(x, y, z)$ on the surface of the star. If we denote its colatitude and its longitude with θ and ϕ, respectively, the Cartesian components of the unit vector $\widehat{OP}$ are: $\widehat{OP} = (\sin\theta\cos\phi, \sin\theta\sin\phi, \cos\theta)$. Note that ϕ is a linear function of the time t because the star is rotating with an angular velocity Ω. If the longitude of P at the time t_0 is ϕ_0, we have: $\phi(t) = \phi_0 + \Omega(t - t_0)$. The Cartesian components of the unit vector pointing towards the observer are: $\hat{s} = (\sin i, 0, \cos i)$, where i is the inclination of the stellar spin axis to the line of sight.

The normal $\hat{n}$ to the stellar surface at the point P is parallel to $\widehat{OP}$ because the star is spherical, that is $\hat{n} = \widehat{OP}$. Therefore, the angle ψ between the normal at P and the direction towards the observer is given by the scalar product $\cos\psi = \hat{n}\cdot\hat{s} = \widehat{OP}\cdot\hat{s}$. Introducing $\mu \equiv \cos\psi$ and substituting the above expressions for $\widehat{OP}$ and $\hat{s}$ into this relationship, we finally obtain:

$$\mu = \sin i \sin\theta \cos[\phi_0 + \Omega(t - t_0)] + \cos i \cos\theta. \qquad (3.1)$$

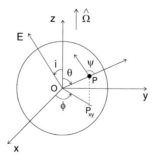

Fig. 3.1 Illustration of the geometry adopted to compute the spot modeling in the case of a single spherical star. The reference frame has its origin at the barycentre O of the star; the z-axis is along the stellar angular velocity Ω; the xy plane coincides with the equatorial plane of the star with the x-axis chosen so that the line of sight is in the xz plane. The spherical coordinates of a point P on the surface of the star are the colatitude θ and the longitude ϕ. The inclination of the stellar rotation axis to the line of sight is i, while ψ is the angle between the normal in P and the line of sight. The projection of the point P on the equatorial plane of the star is indicated with P_{xy} and is introduced to define the longitude ϕ measured with respect to the x-axis

3.2.2   Flux Variation Produced by an Active Region

Let us assume a quadratic limb-darkening law for the unperturbed photosphere in the passband of the observations (cf. Gray 2008):

$$I_u(\mu) = I_0(a + b\mu + c\mu^2),\qquad(3.2)$$

where I_u is the specific intensity in the given passband; I_0 the intensity at the centre of the disc; a, b, and c the limb-darkening coefficients that verify $a + b + c = 1$. The flux emerging from the stellar disc of radius R is:

$$F_u = 2\pi R^2 \int_0^{\pi/2} I_u(\cos\psi)\cos\psi\sin\psi\, d\psi = 2\pi R^2 \int_0^1 I_u(\mu)\mu\, d\mu,\qquad(3.3)$$

where $dA = 2\pi R^2 \sin\theta d\theta$ is the area of the elementary band on the sphere between colatitudes θ and $\theta + d\theta$, and the factor $\cos\psi$ gives its projection on the plane normal to the line of sight. Substituting the limb-darkening law and performing the integration, we find the unperturbed stellar flux:

$$F_u = \pi R^2 I_0 \left(a + \frac{2}{3}b + \frac{1}{2}c\right).\qquad(3.4)$$

In the Sun and sun-like stars, the photospheric active regions are much smaller than the area of the disc. Therefore, we can simplify our treatment by considering point-like active regions. Each one consists of a spotted area A_s and a facular area A_f localized at the same point P with A_s, $A_f \ll \pi R^2$ (cf. Lanza et al. 2003). The flux perturbation produced by that active region, i.e., by its dark spots and bright faculae localized in P, is:

$$\Delta F = \Delta F_s + \Delta F_f = A_s\mu(I_s - I_u) + A_f\mu(I_f - I_u),\qquad(3.5)$$

where I_s is the specific intensity in the spots and I_f that in the faculae. If A is the area of a surface element of the photosphere, we define the filling factor of the spots f_s and that of the faculae Qf according to the relationships:

$$A_s \equiv f_s A, \quad A_f \equiv Qf_s A = QA_s,\qquad(3.6)$$

and their intensity contrasts as:

$$c_s \equiv \left(1 - \frac{I_s}{I_u}\right), \quad c_f \equiv -\left(1 - \frac{I_f}{I_u}\right),\qquad(3.7)$$

where the specific intensity of the spot I_s and of the faculae I_f are given at the same point of the photosphere as the unperturbed intensity I_u. The solar faculae are more contrasted towards the limb and virtually invisible at disc centre. For the sake of simplicity, we assume a linear dependence of their contrast on μ (cf. Lanza et al. 2003, 2004):

$$c_f = c_{f0}(1 - \mu),\qquad(3.8)$$

so that

$$\Delta F = A_s I_u(\mu) \left[-c_s + Q c_{f0}(1 - \mu)\right] \mu = f_s A\, I_u(\mu) \left[-c_s + Q c_{f0}(1 - \mu)\right] \mu.\qquad(3.9)$$

In addition, to further simplify our model, we assume that the spot area A_s and the contrasts c_s and c_{f0} are constant as well as the ratio of the facular-to-spotted area $Q = A_f/A_s$. We also neglect the presence of the spot penumbra as in the first simple models of the variation of the solar irradiance (e.g., Chapman et al. 1984).

The observed flux at the time t is:

$$F(t) = F_u + \Delta F(t),\qquad(3.10)$$

where the time dependence of ΔF comes from the rotation of the star that changes the projection factor μ. Therefore, the relative variation of the flux according to Eq. (3.9) is:

$$\frac{F(t)}{F_u} = 1 + \frac{\Delta F(t)}{F_u} = 1 + \frac{A_s I_u(\mu)}{F_u} \left[Q c_{f0}(1 - \mu) - c_s\right] v(\mu)\mu,\qquad(3.11)$$

or, substituting Eqs. (3.2) and (3.4) into Eq. (3.11):

$$\frac{F(t)}{F_u} = 1 + \left(\frac{A_s}{\pi R^2}\right)\left(\frac{a + b\mu + c\mu^2}{a + 2b/3 + c/2}\right)\left[Q c_{f0}(1 - \mu) - c_s\right] v(\mu)\mu,\qquad(3.12)$$

where the time dependence of the projection factor is given by Eq. (3.1) and v is the *visibility* of the surface element centered at the point P defined as:

$$v(\mu) = \begin{cases} 1 \text{ if } \mu \geq 0 \\ 0 \text{ if } \mu < 0. \end{cases}\qquad(3.13)$$

An illustration of the typical rotational modulation produced by our model active region is given in Fig. 3.2. When the active region is on the invisible hemisphere, the flux is constant at the unperturbed value. When stellar rotation brings the active region into view, the flux initially rises because faculae are more contrasted close to the limb and their dark spots have little effect owing to the foreshortening. As the active region moves towards the centre of the disc, the effect of the faculae becomes less important owing to the decrease of their contrast, while dark spots become dominant as their projected area rises towards disc centre. Finally, when the active

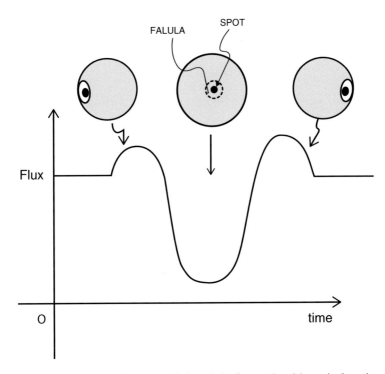

Fig. 3.2 Illustration of the rotational modulation of the flux produced by a single active region consisting of a dark spot and a bright facular area around it with solar-like contrasts

region moves toward the other limb, the flux increases again due to the prevailing effect of the faculae.

When a star is much more active than the Sun, its active regions cannot be treated as point-like features. This is the case of young rapidly rotating stars or of the active components of close binaries that were monitored from the ground thanks to their large light curve amplitudes reaching up to 0.2–0.3 mag in the optical passband (Strassmeier et al. 1997; García-Alvarez et al. 2011). In this case, active regions were generally treated as spherical caps, as discussed in, e.g., Rodonò et al. (1986), Dorren (1987), or Eker (1994). We shall not consider the theory of the light variations produced by such extended spots, referring the interested reader to those works and the references therein. However, a few results obtained with those models will be mentioned in Sect. 3.2.3.

An important geometrical parameter affecting starspot modelling is the inclination of the stellar spin axis i. If the photometric period P_{rot} is known from timeseries photometry, the rotational broadening of the spectral lines $v \sin i$ is measured from high-resolution spectroscopy, and the radius of the star R is estimated from models or interferometry, we can derive the inclination from: $\sin i = P_{rot}(v \sin i)/2\pi R$. This method can be applied to young rapidly rotating stars because for stars similar to the Sun the relative error on $v \sin i$ is of ≈ 50–100% due to the effects of macroturbulence, even when very high-resolution spectra are available. Moreover, for those rapidly rotating stars, the inclination can be derived by minimizing systematic errors in the process of constructing Doppler imaging maps (e.g., Rice and Strassmeier 2000). For stars that rotate slowly, the inclination is generally unknown or can be estimated with large uncertainties. An intermediate case is that of stars that rotate with $\Omega \geq (2-3)\Omega_\odot$ for which asteroseismology can be applied to derive the inclination because the visibility of the p-modes belonging to a given rotationally split multiplet, that differ by the azimuthal order m, depends on the inclination (e.g., Ballot et al. 2006, 2011).

The role of faculae is parametrized by Q in our simple model. In general, light variations in stars remarkably more active than the Sun seem to be dominated by dark spots (Lockwood et al. 2007) and also in the Sun the relative contribution of the faculae decreases during the maximum phase of the 11-year cycle (e.g. Foukal 1998). The recent works by Gondoin (2008) and Messina (2008) provide more information on the facular contribution. A method to estimate Q from single-band light curves, thanks to the different shapes of the facular and spot light modulations, was introduced by Lanza et al. (2003), and its applications are mentioned in Sect. 3.2.3.

We have considered the case of photometry in a single passband because space-borne telescopes generally observe in a single wide passband to maximize the flux and reduce the photon-shot noise when searching for planetary transits, or have a few non-standardized passbands, such as CoRoT (Auvergne et al. 2009). However, ground-based photometry is often acquired in several standard photometric passbands that allow to estimate the spot temperature (e.g., Poe and Eaton 1985; Strassmeier and Olah 1992). For this application, it is important an appropriate modelling of the limb-darkening in the different passbands. In Fig. 3.3, the synthetic light curves produced by a starspot on a solar-like star in two different passbands are plotted. The spot is completely dark, that is no flux is coming from it. Therefore, the observed colour variation is due solely to the different limb-darkening coefficients in the two passbands and amounts to $\approx 10\%$ of the amplitude of the light modulation. Therefore, a word of caution is in order when interpreting colour modulations as immediate proxies for starspot temperature. If the intrinsic flux of the spot in the considered passbands is low, as it is often the case in the U or B passbands, the colour variation can be dominated by differential limb darkening rather than by the spot temperature deficit.

3.2.3 Few-Spot Models

The light curves of a spotted star are generally not sinusoidal, therefore a single spot is not enough to obtain an adequate fit. The simplest models consider two or three non-overlapping spots. In order to compute those models, it suffices to add the effects of individual active regions as introduced in Sect. 3.2.2. For the case of two spots, the free parameters are: the inclination i, the rotation period $P_{rot} = 2\pi/\Omega$,

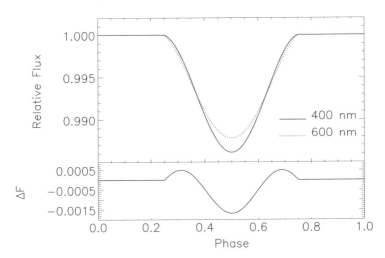

Fig. 3.3 *Upper panel*: Synthetic flux vs. rotation phase at two wavelengths (400 and 600 nm) in the case of a completely dark starspot transiting across the disc of a late-type star. The two *light curves* are different because of the dependence of the limb-darkening coefficients on the wavelength. *Lower panel*: the flux difference (*colour*) vs. the rotation phase

the limb-darkening coefficients, the unperturbed flux level F_u, the spot and facular contrasts c_s and c_{f0}, the ratio of the facular-to-spotted areas Q; and, for each spot, the relative area $A_s/\pi R^2$, the colatitude θ, and the initial longitude ϕ_0. The unperturbed level is generally unknown, so it is usually fixed at the maximum observed flux or allowed to vary by 0.1–1 % above that level because this can sometimes improve the best fit by providing the model with an additional degree of freedom that allows it to converge to a deeper minimum in the χ^2 landscape. The parameters i, P_{rot}, and Q are generally fixed and only the *geometrical parameters*, i.e., A_s, θ, and ϕ_0 for each of the spots, are varied to minimize the χ^2 of the model. Sometimes, a flux term independent of the spot longitude is added to the model to account for a uniformly distributed pattern of spots that does not produce any flux modulation, but affects the mean light level (cf. Lanza et al. 2003).

In some cases, thanks to the small number of varied parameters, the model may be unique, for example, a well-defined minimum of the χ^2 can be found in the six-parameter space of a two-spot model. However, in most of the cases, degeneracies among the parameters are present, especially when the accuracy of the photometry is limited. This can be understood if one considers for simplicity a model with only one dark spot without faculae ($Q = 0$). The minimum of light corresponds to the transit of the spot on the central meridian of the star's disc and it allows us to derive the initial longitude of the spot ϕ_0. The minimum of light occurs at time t_m when $(t_m - t_0) = -\phi_0/\Omega$ and its amplitude is: $|\Delta F|_{max}/F_u = c_s(A_s/\pi R^2)\cos(i - \theta)$, where limb-darkening is neglected for simplicity. The duration of the spot transit δt corresponds to the two longitudes where $\mu = 0$, i.e., $\delta t = t_2 - t_1$, where $\sin i \sin \theta \cos[\phi_0 + \Omega(t_k - t_0)] + \cos i \cos \theta = 0$, with $k = 1, 2$. Therefore, if the inclination $i \neq 90°$, the unspotted flux level F_u, and the other physical parameters are known, we can derive the three geometrical parameters A_s, θ, and ϕ_0 of our single spot from the duration δt, the time of light minimum t_m, and the amplitude of the light minimum $|\Delta F|_{max}/F_u$. If the inclination is $90°$, the duration δt of the transit of the spot across the disc becomes independent of its colatitude and we loose the information on that parameter. Even if $i \neq 90°$, the finite precision of the photometry introduces uncertainties on the spot location and area. If the inclination i is not known a priori, the colatitude θ, the unprojected area A_s, and the inclination i itself become largely degenerate because the combination $A_s \cos(i - \theta)$ appears in the relationship for the amplitude of the flux modulation. These considerations show that spot modelling can give unique solutions only in very special cases.

When we fit two spots and the inclination is only poorly estimated, we expect strong degeneracies to arise among the different parameters because different combinations of the individual spot areas and colatitudes give similar light modulations, especially when the inclination is close to $90°$. In spite of such limitations, two-spot models have been widely applied to fit ground-based photometry for which the precision is of the order of 0.01 mag. To obtain a sufficient coverage in phase, the data gathered along an entire season were generally used to construct an average light curve, thus averaging short-term changes in the spot pattern. Rodonò et al. (1986) provided some examples of that kind of spot models, generally yielding a well-defined minimum in the χ^2 space thanks to the limited number of free parameters. Spots at high latitudes and even at the poles were often found because the model used them to adjust the variations of the mean light level given that they were circumpolar for inclination $i \neq 90°$ and therefore always in view.

Recently, two-spot models have been resumed and applied within a Bayesian framework using Monte Carlo Markov Chain techniques to fully explore the *a posteriori* parameter distributions and their degeneracies (Croll 2006; Fröhlich 2007; Lanza et al. 2014). The applications were focussed on estimating the amplitude of the surface differential rotation by allowing the two spots to have different rotation periods. A two-spot model with non-evolving spots was applied to individual time intervals sufficiently short to avoid that the intrinsic spot evolution affects the result.

Other applications of models with a few spots are the estimation of the facular-to-spotted area ratio Q or of the maximum time interval during which the spot pattern is unaffected by the intrinsic starspot evolution. Lanza et al. (2003) first performed those applications for the Sun and then for some CoRoT and Kepler targets (cf. Lanza et al. 2009a,b, 2010, 2011; Bonomo and Lanza 2012). Note that, since Q appears in combination with the facular contrast parameter c_{f0} in the product $Q c_{f0}$ (cf. Eq. (3.12)), it is possible to determine Q only by fixing c_{f0}.

3.2.4   Multispot Models with Evolution

The advent of automated photometric telescopes in the 1990s allowed to follow the evolution of the light modulation of active stars in a systematic way and posed the problem of modelling their spot evolution (e.g. Rodonò et al. 2001). Strassmeier and Bopp (1992) were among the first to propose a model that incorporated the intrinsic evolution of the starspots and the relative drift in longitude owing to surface differential rotation. With space-borne telescopes such as CoRoT, multi-spot models with evolution became even more important. Mosser et al. (2009) fitted the light curves of several CoRoT asteroseismic targets by applying a model with evolving spots, usually limited to 2–3 per stellar rotation. Best fits were obtained with an extended exploration of the geometric parameter space by means of a relaxed χ^2 minimization based on a technique similar to simulated annealing. Their method was extensively tested with simulated data and compared to other approaches to study the dependence of the results on model assumptions and on the parameters held fixed. The model proved useful to derive robust estimates of the spot lifetimes and mean rotation period, while other parameters, such as the inclination of the spin axis (independently known from asteroseismology), spot latitudes, and differential rotation were found sensitive to model assumptions.

Frasca et al. (2011) and Fröhlich et al. (2012) (cf. also Fröhlich et al. 2009), applied multispot models with up to 7–9 evolving spots to fit Kepler timeseries of several hundred days. A Bayesian approach was used to derive the *a posteriori* free parameter distributions generally including the inclination and the surface differential rotation.

The main limitation of multi-spot models, in addition to the strong parameter degeneracies, is the large amplitude of the residuals in comparison with the photometric errors. This is especially critical when we model the light curves of eclipsing binaries because the eclipse profile is highly sensitive to the shape and location of the occulted spots. For these reasons, continuous spot models, similar to those considered for Doppler imaging (Vogt et al. 1987), have been introduced since the second half of the 1990s to improve the best fits of the light curves. They will be the subject of the next sections.

3.3   Models with Continuous Spot Distributions

A continuous distribution of spots on the surface of a star can be specified by giving the spot filling factor f_s in each surface element. We define $f_s \equiv A_s/A$, where A_s is the spotted area within the surface element of area A, as considered above. The spot distribution is mapped by the distribution of f_s over the surface of the star. Since the light curve is a one-dimensional dataset, while the filling factor map is a two-dimensional function, i.e., $f_s = f_s(\theta, \phi)$, the problem of finding f_s given the light curve has generally many different solutions and the map is also highly sensitive to small variations in the input dataset. In the mathematical language, this is a *ill-posed problem* (cf. Tikhonov and Goncharsky 1987).

The usual method to solve this kind of problems is by combining the information coming from the light curve with some a priori information in order to obtain a unique and stable solution, i.e., a map that does not vary greatly when there are small variations in the dataset, or, in other words, that is not critically sensitive to the effect of the errors in the photometry. A simple way of introducing a priori assumptions in the solution process is by restricting the shape and number of the spots, as we did in the previous discrete spot models. A more sophisticated way is that of coding some statistical property that we want to impose to the solution into an appropriate functional. This is the method of *solution regularization* that will be described below. However, before introducing the mathematical formulation of regularization, we need to compute the flux emergent from the stellar disc in the presence of a continuous spot distribution.

3.3.1 Flux Variation Produced by a Continuous Distribution of Active Regions

For simplicity, we subdivide the surface of the star into a large number of elements N, each of area A_k, where $k = 1, \ldots, N$. The flux coming from the kth element is:

$$\delta F_k = I(\mu_k) A_k \mu_k v(\mu_k), \tag{3.14}$$

where

$$I(\mu_k) = f_s I_s + Q f_s I_f + [1 - (Q + 1) f_s] I_u(\mu_k). \tag{3.15}$$

This equation gives the average specific intensity emerging from the given surface element as the result of the intensity coming from the spotted photosphere with a filling factor f_s, from the facular photosphere with a filling factor $Q f_s$, and from the unperturbed photosphere, the filling factor of which is $1 - (Q + 1) f_s$. With little algebra, we find:

$$I(\mu_k) = \left\{ 1 + \left[c_{f0} Q (1 - \mu_k) - c_s \right] f_s \right\} I_u(\mu_k). \tag{3.16}$$

The total flux coming from the disc is:

$$F(t) = \sum_{k=1}^{N} \delta F_k = \sum_k A_k I_u(\mu_k) \left\{ 1 + \left[c_{f0} Q (1 - \mu_k) - c_s \right] f_k \right\} v(\mu_k) \mu_k, \tag{3.17}$$

where f_k is the spot filling factor (previously indicated with f_s), μ_k the projection factor of the kth surface element at the time t (cf. Eq. (3.1)), and $v(\mu_k)$ the visibility function in Eq. (3.13).

In general, we want to compute M flux values $F_j \equiv F(t_j)$, where t_j are the times of the observations, with $j = 1, \ldots, M$; we define them as the model flux vector $\mathbf{F} \equiv \{F(t_j), j = 1, \ldots, M\}$. We can express its relationship to the distribution of the filling factor on the surface of the star by introducing an $M \times N$ projection matrix $\tilde{\mathbf{R}} = \{R_{jk}\}$ and a constant vector C_u such as:

$$F_j \equiv F(t_j) = \sum_k R_{jk} f_k + C_{uj}, \tag{3.18}$$

or, in matrix notation:

$$\mathbf{F} = \tilde{\mathbf{R}} \mathbf{f} + \mathbf{C}_u, \tag{3.19}$$

where $\mathbf{f} = \{f_k, \ k = 1, \ldots, N\}$ is the vector of the filling factor on the surface of the star and

$$R_{jk} \equiv A_k I_u(\mu_k) \left[c_{f0} Q(1 - \mu_k) - c_s \right] v(\mu_k) \mu_k \ \text{ with } \mu_k = \mu_k(t_j), \tag{3.20}$$

and

$$C_{uj} \equiv A_k I_u(\mu_k) v(\mu_k) \mu_k = F_{uj}, \tag{3.21}$$

(cf. (3.17)) where we introduce the vector of the unperturbed flux $\mathbf{F}_u$ consisting of M constant components, i.e., $\mathbf{F}_u \equiv \{F_{uj}, j = 1, \ldots, M\}$ with $F_{uj} = F_u$.

3.3.2 The Light Curve Inversion Problem and the Regularization

We now consider the inverse problem of deriving the distribution of the spot filling factor from the light curve dataset. If the observed flux values at the times t_j are denoted as the vector $\mathbf{D} = \{D_j, j = 1, \ldots, M\}$, we can first consider the ideal case when: (a) there are no measurement errors; (b) our model for the flux variations is exact; and (c) the unspotted flux is known. In this case, one may hope to derive a solution for the filling factor vector $\mathbf{f}$, by solving the linear system:

$$\tilde{\mathbf{R}} \mathbf{f} = \mathbf{D} - \mathbf{C}_u. \tag{3.22}$$

In general, this system has infinite solutions because the matrix $\tilde{\mathbf{R}}$ is singular, i.e., it has a nullspace of finite dimension whose vectors $\mathbf{f}_{null}$ have the property $\tilde{\mathbf{R}} \mathbf{f}_{null} = 0$

(see Press et al. 2002, Ch. 2). Therefore, if a given vector $\mathbf{f}_0$ is a solution of Eq. (3.22), $\mathbf{f}_0 + h\mathbf{f}_{null}$, where h is any real number, is a solution too. Cowan et al. (2013) investigated the nullspace in some light curve inversion problems showing that it can significantly affect the solution. From a geometrical point of view, the existence of the null space is associated with particular distributions of brightness on the stellar surface that do not produce a light modulation as the star rotates (see Cowan et al. 2013, for some examples).

A more realistic case is that of a dataset with finite errors. In this case, we look for a solution that minimizes the χ^2 between the dataset and the model. Specifically, the χ^2 corresponding to a given distribution of the filling factor is:

$$\chi^2(\mathbf{f}) \equiv \sum_{j=1}^{M} \frac{(D_j - F_j)^2}{\sigma_j^2}, \qquad (3.23)$$

where σ_j is the standard deviation of the flux measurement D_j.

In general, the solution found by minimizing the χ^2 is not unique and is highly sensitive to small changes in the dataset, in the sense that a small change in the data vector $\mathbf{D}$ produces a large change in the filling factor distribution $\mathbf{f}$. The idea of regularization is to add to the χ^2 an appropriate mathematical function of $\mathbf{f}$ that warrants a unique and stable solution, i.e., a solution that varies in a continuous way in the $\mathbf{f}$ space. Note that in general the filling factor is a function $f(\theta, \phi)$ that we have discretized into a vector $\mathbf{f}$ of N elements, therefore the regularizing term is, mathematically speaking, a functional. There are several possible choices that have been investigated by the mathematicians and proved effective in our kind of inversion problem (cf. Tikhonov and Goncharsky 1987; Titterington 1985).

The most widely used is the maximum entropy functional that provides a quantitative measure of the configuration entropy of the map, i.e., of the information necessary to transmit the map expressed as a sequence of bytes (Bryan and Skilling 1980; Narayan and Nityananda 1986). It assumes a default map as a reference and measures the difference in the information content of the considered map with respect to it. In our case, the default map corresponds to an immaculate star. The specific formulation of the maximum entropy functional that I prefer is that given by Collier Cameron (1992).

The regularized solution is computed by minimizing an objective function Z defined as a linear combination of the χ^2 and the regularizing functional S. For the maximum entropy case:

$$Z(\mathbf{f}) = \chi^2(\mathbf{f}) - \lambda_{ME} S(\mathbf{f}), \qquad (3.24)$$

where $\mathbf{f} = \{f_k, \ k = 1, \ldots N\}$ is the vector of the spot filling factors for the individual surface elements, $\lambda_{ME} > 0$ a Lagrangian multiplier, and

$$S = -\sum_{k} w_k \left[f_k \log \frac{f_k}{m} + (1 - f_k) \log \frac{(1 - f_k)}{(1 - m)} \right], \qquad (3.25)$$

Fig. 3.4 The χ^2 landscape in the case of a simple model with only two surface elements $\mathbf{f} = \{f_1, f_2\}$, showing the minimum of the χ^2, i.e., χ_0^2, and the effect of the regularization ($\lambda_{\mathrm{ME}} > 0$) that increases the χ^2 value, driving at the same time the solution towards the unspotted map with $\mathbf{f} = 0$. The effect of a different regularizing functional that moves the solution along a different path is also shown for comparison

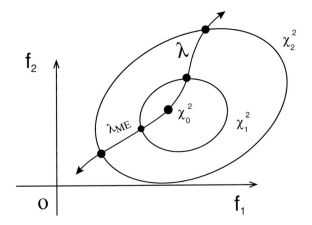

is the entropy functional, where w_k is the relative area of the kth surface element and $m = 10^{-6}$ is a default minimum spot filling factor included to avoid the divergence of the logarithm. S gets its maximum value equal to zero for an immaculate star, i.e., $f_k = m$ in each surface elements.

The effect of the regularization is that of reducing the spot filling factor (or the spotted area) as much as possible, compatibly with fitting the data, by increasing the Lagrangian multiplier. In Fig. 3.4, we show the isocontours of the χ^2 landscape for the illustrative case of a map consisting of only two surface elements that we use to explain the concept. Without regularization (i.e., $\lambda_{\mathrm{ME}} = 0$), the best fit has the minimum $\chi^2 = \chi_0^2$ and the residuals of the fit have a Gaussian distribution with a mean value $\mu = 0$ and a standard deviation σ_0. In general, the best fit with $\chi^2 = \chi_0^2$ is not acceptable because we fit also some component of the measurement errors and the solution is unstable. With the regularization, ($\lambda_{\mathrm{ME}} > 0$), the fit has $\chi^2 = \chi_1^2 > \chi_0^2$ and the residual distribution is now centred at a value $\mu < 0$ because the spotted area is reduced. However, the solution becomes stable and unique for a sufficiently large value of λ_{ME}.

The role of the a priori information introduced through the regularization is that of selecting one specific solution vector $\mathbf{f}$ among the infinite ones that correspond to the condition $\chi^2 = \chi_1^2$. In the case of the maximum entropy solution, the selected vector $\mathbf{f}$ corresponds to the solution that minimizes the individual f_k, while verifying the condition $\chi^2 = \chi_1^2$. Of course, it is possible to move along a different line in the χ^2 landscape which corresponds to a different kind of regularization. The fundamental requisite for the choice of the regularizing functional is that it must lead to a unique and stable solution when the Lagrangian multiplier λ is sufficiently large.

In the case of the maximum entropy regularization, we fix the optimal value of λ_{ME} by comparing μ with σ_0, the standard deviation of the residuals as obtained with the unregularized best fit (i.e., for $\lambda_{\mathrm{ME}} = 0$). The signal-to-noise ratio of a light curve can be defined as $S/N = \mathcal{A}_{\max}/\sigma_0$, where $\mathcal{A}_{\max}$ is the maximum amplitude of the light modulation due to the starspots. By increasing λ_{ME}, the fit is shifted

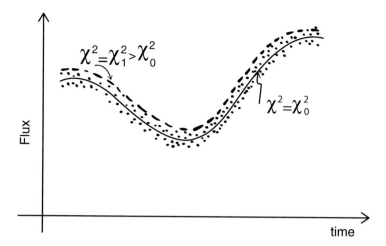

Fig. 3.5 Illustration of the best fit of a light curve without regularization (*solid line*) with $\chi^2 = \chi_0^2$ and with regularization having $\chi^2 = \chi_1^2 > \chi_0^2$ (*dashed line*). The latter is higher than the best fit corresponding to the minimum of the $\chi^2 = \chi_0^2$ and its residuals are not symmetrically centred on the zero value because the regularization smooths out the spot pattern driving the solution towards the unspotted flux level, here assumed to be higher than the light maximum

towards the unspotted level (see Fig. 3.5), while the distribution of the residuals is shifted towards negative values and its standard deviation σ increases because the regularization smooths out the small spots that were previously used to fit the noise components and reduce the χ^2 (see Fig. 3.6). A practical recipe to fix λ_{ME} in the case of photometry with high signal-to-noise ($S/N \geq 100$) consists in increasing λ_{ME} until:

$$\mu = \frac{\sigma_0}{\sqrt{M}}, \tag{3.26}$$

where M is the number of data points in the light curve (cf. Lanza et al. 2009a). When $S/N \approx 10$–30, we need to adopt a stronger regularization to reduce the impact of the noise, i.e.:

$$\mu = \beta \frac{\sigma_0}{\sqrt{M}}, \tag{3.27}$$

where $1.5 \leq \beta \leq 3$ is a numerical factor (cf. Lanza et al. 2009b). A visual inspection of the fit is generally needed to find the largest possible acceptable deviations, i.e., to fix the appropriate value of β by considering the trade-off between the accuracy of the fit and its smoothness.

Another regularizing functional often adopted is the Tikhonov functional T. It selects the smoothest map compatible with the data, i.e., the one that minimizes the average $|\nabla f(\theta, \phi)|^2$ over the stellar surface (Piskunov et al. 1990). In other words,

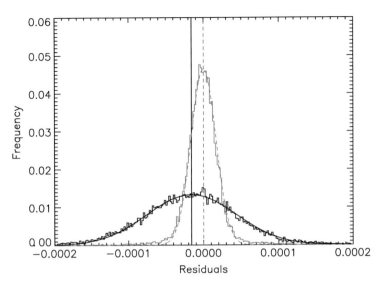

Fig. 3.6 Distribution of the residuals of the best fit of a light curve obtained without any regularization (*green histogram*) together with its Gaussian best fit (*green dashed line*) centred at zero value, as indicated by the *dashed green vertical line*. The distribution of the residuals after applying the maximum entropy regularization is shown by the *black histogram* together with its Gaussian best fit (*black solid line*); the *solid vertical line* marks the mean of the Gaussian best fit. Note that the distribution of the residuals of the regularized solution is centred at a negative value because the corresponding fit is systematically higher than the photometric data points as shown in Fig. 3.5. Its standard deviation is greater than that of the unregularized solution because the spot pattern is smoother owing to the regularization

one seeks to minimize a linear combination $Z = \chi^2 + \lambda_T T$, where $\lambda_T > 0$ is the Lagrangian multiplier and:

$$T(\mathbf{f}) = \int_\Sigma \left[\left(\frac{\partial f}{\partial \theta} \right)^2 + \frac{1}{\sin^2 \theta} \left(\frac{\partial f}{\partial \phi} \right)^2 \right] d\Sigma, \tag{3.28}$$

where Σ is the surface of the star whose element is $d\Sigma = \sin\theta d\theta d\phi$. Of course, other regularizing functionals are possible, e.g., that introduced by Harmon and Crews (2000) and applied by Roettenbacher et al. (2013).

A crucial limitation of spot modelling is that we use one-dimensional information, that is, a light curve, to reconstruct a two-dimensional map of the stellar surface. Most of the applications of regularized spot modelling have targeted close eclipsing binaries or, more recently, active stars with transiting planets whose inclination is close to 90°. Therefore, the information on spot latitude is very limited or non-existent. In those cases, it is better to collapse the two-dimensional map obtained by a regularized model into a one-dimensional distribution of the spot filling factor versus the longitude and consider that distribution as the final product of the modelling. The *relative variation* of the spotted area vs. the longitude has little dependence on the specific regularization adopted and can be considered as a robust result of the analysis (cf. Lanza et al. 1998, 2006). In other words, the absolute value of the spotted area depends on the often unknown spot contrast and unspotted light level, but its relative distribution vs. longitude can be derived thanks to the light modulation that it produces as the star rotates. Similarly, the long-term variations in the relative spotted area can be considered a robust result of the modelling, if we assume that the spot contrast stays constant in the given passband. This allows us to detect stellar activity cycles akin to the 11-year sunspot cycle.

3.3.3 Alternative Approaches

The minimization of the χ^2 can be approached also by means of the *singular value decomposition* (hereafter SVD) of the projection matrix $\tilde{\mathbf{R}}$. The method is described in, e.g., Press et al. (2002). The main advantage is that the linear combinations of the components of $\mathbf{f}$ that are not constrained by the data are driven to zero or to small, insignificant values, while the solution becomes dominated by the linear combinations of the elements of $\mathbf{f}$ that can account for most of the flux variation. These are the so-called *principal components*. The number of components retained in the solution is determined by the minimum acceptable singular value. An advantage of the method is that the errors of the individual components can be evaluated starting from the errors of the individual photometric data. In the case of the regularized models, the statistical errors on the f_k are not easily estimated because the a priori information introduced into the solution usually dominates. Therefore, systematic errors can be larger than the statistical errors in most of the cases and only a comparison between maps obtained with different regularizing functionals provides some insight into the errors (Lanza et al. 1998).

Several spot modelling approaches based on the general principle of SVD or principal component analysis have been proposed, e.g., by Berdyugina (1998) or Savanov and Strassmeier (2005, 2008) who also performed comparisons with test cases and studied the general properties of the solutions.

Finally, it is worth mentioning the approach by Cowan et al. (2013) who applied a Fourier decomposition method to extract a map from the observed rotational flux modulation. In principle, all the Fourier components of the spot map characterized by different azimuthal orders m can be extracted in the case of an ideal noiseless light curve sampled with perfect continuity. In practice, their amplitude decreases as $|m^2 - 1|^{-1}$ for $m > 1$. This implies that in the case of a real light curve, the amplitudes of the higher order Fourier components become soon comparable with or smaller than the noise, making it impossible to accurately extract them. In other words, it becomes impossible to resolve sufficiently localized brightness inhomogeneities. For this reason, a model based on a discrete (or continuous) spot distribution is generally superior to Fourier decomposition in the case of active stars.

3.4 Spot Occultations During Planetary Transits

The observations of extrasolar planets transiting their host stars opened a new avenue in the investigation of other planetary systems. Here, I shall consider only the contribution that planetary transits give to the modelling of the distribution of the surface brightness on the disc of their host stars. In Fig. 3.7, the case of a planet transiting across the disc of a star with a dark spot along the occulted band is shown, neglecting for simplicity the effects of the limb darkening. When the planet's disc is not covering the spot, the flux is reduced by the spot, but the variation of the flux vs. the time has exactly the same shape as when the spot is not on the stellar disc. However, when the spot is occulted by the planet, the flux shows a relative increase because the configuration corresponds to that of a planet transiting across the disc of a star without spots, that is, whose flux is higher. The position of the centre of the light bump gives a measure of the spot longitude on the stellar disc, while its extension gives a measure of the size of the spot, or, to be precise, of the extension in longitude of the portion of the spot that is occulted by the planet (e.g., Wolter et al. 2009). Finally, the amplitude of the bump depends on the contrast of the spot that can provide a measure of its temperature when the effective temperature of the unperturbed photosphere is known. Silva-Valio et al. (2010) determined the distributions of the size and contrast of the spots occulted in CoRoT-2 using this approach. This method is unique to resolve small spots ($\sim$50 Mm) on slowly

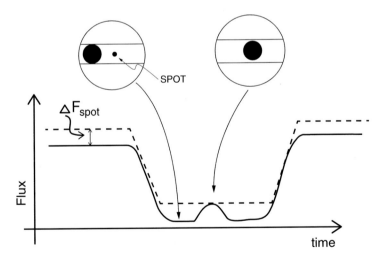

Fig. 3.7 The case of a transiting planet occulting a *dark spot* (at the disc centre) along the transit chord. The flux variation due to the transit in the case of a star with the spot is given by the *solid line* in the lower plot that shows a relative flux increase (a bump) when the spot is occulted. This happens because the corresponding configuration is the same as in the case without spot. In that case the reference flux level outside the transit is higher as shown by the *dashed line* that is the flux variation vs. the time in the case of a star without spots. The quantity ΔF_{spot} measures the flux decrease due to the spot when it is in view on the stellar disc

rotating, sun-like stars that cannot be mapped through Doppler imaging techniques. In principle, spots as small as a few Mm can be resolved if an Earth-size planet occults them, although the photometric accuracy of CoRoT and Kepler is generally insufficient to do that on individual transits (cf. Barros et al. 2014).

Starspot occultations can be used to derive the rotation rate of the star in the latitude band occulted by the planet, if the orbit of the planet and the equator of the star are aligned, because successive occultations of the same spot along successive transits can be used to precisely measure the variation of its longitude vs. the time (Silva-Valio 2008). On the other hand, if the stellar spin and the orbital angular momentum are not aligned, the planet cannot occult the same spot along successive transits because the rotational motion of the spot across the stellar disc is not parallel to the transit chord (Nutzman et al. 2011; Sanchis-Ojeda and Winn 2011; Sanchis-Ojeda et al. 2013). Therefore, monitoring starspot occultations along successive transits can provide information on the projected alignment of the stellar spin and orbital angular momentum in a planetary system. Similar information can be obtained through the observation of the radial-velocity anomaly induced by the transit, i.e., the so-called Rossiter–McLaughlin effect, that allows a measurement of the projected misalignment, although limited to stars with a $v \sin i \geq 2 - 3 \, \mathrm{km \, s^{-1}}$ (e.g., Albrecht et al. 2012). Note that different models of planetary system formation predict different misalignment distributions, therefore such a kind of measurements provides stringent tests to those models.

The precise timing of planetary transits should also take into account the distortions of the transit profile due to spot occultations. This can be a subtle effect when the photometric accuracy is not high enough to resolve the individual bumps (Oshagh et al. 2013; Barros et al. 2014).

3.5 Tests of Spot Models

Many tests of spot models have been published in the literature since the beginning of their application. Among the classic works, I refer to Kovari and Bartus (1997) for two-spot models. Here, I shall consider only a few tests that are based on a comparison with direct observations in the case of the Sun or with the results obtained with independent methods such as Doppler imaging or starspot occultations during planetary transits.

The total solar irradiance (hereafter TSI) provides a good proxy for the modulation of the Sun as a star because its variation is dominated by photospheric sunspots and faculae that produce most of their effects in the optical passband, although the relative variations becomes larger and larger at shorter and shorter wavelengths, in particular if we consider those associated with the solar cycle. Our star is seen almost equator-on, therefore the TSI light curve does not contain information on the latitudes of the active regions, but only on their longitudes and area variations. Lanza et al. (2007) performed detailed tests for different spot modelling approaches by fitting the TSI modulation over an extended portion of solar cycle 23. They applied three-spot, maximum entropy, and Tikhonov regularized models, and compared the derived distributions of the filling factor vs. longitude with those of observed sunspot groups as well as with the variation of their total area. Adopting a fixed value Q for the facular-to-spotted area ratio, they found a remarkably good reproduction of the longitude distribution of the sunspot groups during the rising and the maximum phase of the cycle with the maximum entropy model performing significantly better than the three-spot and Tikhonov models. The resolution in longitude of the models was about $40°$–$50°$ considering a typical $S/N \sim 50$–100.

The facular component in the active regions can induce systematic shifts in their derived longitudinal distribution because faculae have a photometric effect that reaches the maximum close to the limb, while dark spots produce their maximum effect closer to the disc centre. Therefore, the model can shift the longitude of a given active region in an attempt to better reproduce the light modulation with the constraint of a fixed Q. The variation in the total spotted area is also systematically affected by the value of Q, but the overall variations due to the solar 11-year cycle are reproduced, in particular by the maximum entropy models.

Lanza et al. (2007) conclude that the maximum entropy model provides the most accurate description of the distribution of the active regions vs. longitude in the Sun, in particular when the $Q = 9$ value they adopted is the most appropriate, that is in the rising and maximum phases of the 11-year cycle. The good reproduction of the overall variations in the total sunspot area supports the use of that spot modelling to detect stellar activity cycles. The reason why the maximum entropy model is better in comparison to the discrete and Tikhonov models is probably associated with the low level of activity of our star that is characterized by several small active regions simultaneously present on the stellar disc. The three spot models has too few degrees of freedom to account for the complexity of the pattern, especially during the rising and the maximum cycle phases, while the Tikhonov maps display too smooth and extended features that are not observed on the Sun.

In the case of distant stars, the results of Lanza et al. (2007) support the use of spot modelling to derive active spot longitudes and activity cycles. An independent test by comparing maximum entropy models with an extended sequence of Doppler imaging maps was performed by Lanza et al. (2006) in the case of the highly active close binary HR 1099 for which long-term photometry from the ground was available. The results support the possibility of deriving the distribution of the starspots vs. the longitude, although with a limited resolution ($\approx 100°$) because of an $S/N \sim 10$–30 attainable from the ground.

Another interesting test was performed by Silva-Valio and Lanza (2011) in the case of the planetary host CoRoT-2. From the out-of-transit light curve, Lanza et al. (2009a) computed a maximum entropy spot model that provided them with the distribution of the spot filling factor vs. the longitude and time. It was compared with the longitudes of the spots occulted during the transits finding a remarkably good agreement. Although planetary occultations provided a significantly higher longitude resolution, the locations of the active longitudes where starspots were preferentially found were reproduced very well and also their migration vs. the time was very similar.

CoRoT-2 became also a benchmark to test different spot modelling approaches. For example, Huber et al. (2010) considered a model in which the surface of the star was subdivided into 12 non-occulted sectors and 24 sectors along the occulted chord, varying their brightness to fit the light curve. They obtained a spot map remarkably similar to that of Lanza et al. (2009a) that was based on the out-of-transit light curve only, thus confirming their results. Another test came from the comparison with a Bayesian few-spot model by Fröhlich et al. (2009).

Independent confirmations are particular important in view of the results on the active longitudes, spot lifetimes, surface differential rotation, and short-term activity cycles obtained for CoRoT-2 as well as for other stars with close-in transiting planets such as Kepler-17 (Bonomo and Lanza 2012). For a detailed discussions of these topics, I refer to the cited original papers and to Lanza (2015) for the possibility of star-planet interactions affecting stellar photospheric activity.

3.6 Conclusions

I briefly reviewed the foundations of spot modelling, the relevance of which is becoming increasingly greater thanks to the availability of high-precision, high-duty cycle light curves acquired by space-borne telescopes designed to look for transiting planets around solar-like stars (cf. Rauer et al. 2014). Different mapping techniques can be applied to derive the distribution of the spotted area vs. longitude and its relative time variation with good confidence, especially in the case of stars with transiting planets for which the inclination can be safely derived or reasonably guessed from the measurement of their projected spin-orbit angle. CoRoT-2 is a benchmark case for the comparison of different modelling approaches as well as for the phenomena that can be detected with spot modelling such as active longitudes, spot evolution, differential rotation, and short-term activity cycles (Lanza et al. 2009a).

I limited myself to the standard spot models proposed for active stars. However, specialized models for stars with transiting and non-transiting planets are expected to become even more important in the near future because they provide information on the spin-orbit alignment of the systems. The radial velocity jitter associated with stellar active regions is a major limitation to the detection and measurement of the mass of Earth-sized planets (e.g. Haywood et al. 2014). Spot modelling techniques can be applied to mitigate its impact as shown by recent investigations (cf. Dumusque 2014; Dumusque et al. 2014, and references therein).

The possibility of extending spot modelling to pre-main sequence stars is also interesting, although limited to those objects the light variations of which are dominated by photospheric brightness inhomogeneities (Cody et al. 2014). Finally, a word of caution is in order in the case of close binary systems where the light modulations due to different effects, such as ellipsoidicity, reflection, gravity darkening, and Doppler beaming can combine with those due to surface brightness inhomogeneities to produce a complex phenomenology the modelling of which is a very challenging task (cf. Lanza et al. 1998; Kallrath and Milone 1999; Herrero et al. 2013, 2014).

Acknowledgements I am grateful to the organizers of the Besançon School on Stellar Cartography, Prof. J.-P. Rozelot and Dr. C. Neiner, for their kind invitation and for fostering a lively and stimulating environment during the school. I enjoyed discussions with several lecturers and participants that helped me a lot to improve my understanding of the different methods for stellar mapping and their applications.

References

Albrecht, S., Winn, J. N., Johnson, J. A., Howard, A. W., Marcy, G. W., Butler, R. P., et al. (2012). Obliquities of hot Jupiter host stars: Evidence for tidal interactions and primordial misalignments. *The Astrophysical Journal, 757*, 18.

Auvergne, M., Bodin, P., Boisnard, L., Buey, J.-T., Chaintreuil, S., Epstein, G., et al. (2009). The CoRoT satellite in flight: Description and performance. *Astronomy & Astrophysics, 506*, 411.

Ballot, J., García, R. A., & Lambert, P. (2006). Rotation speed and stellar axis inclination from p modes: How CoRoT would see other suns. *Monthly Notices of the Royal Astronomical Society, 369*, 1281.

Ballot, J., Gizon, L., Samadi, R., Vauclair, G., Benomar, O., Bruntt, H., et al. (2011). Accurate p-mode measurements of the G0V metal-rich CoRoT target HD 52265. *Astronomy & Astrophysics, 530*, A97.

Barros, S. C. C., Almenara, J. M., Deleuil, M., Diaz, R. F., Csizmadia, S., Cabrera, J., et al. (2014). Revisiting the transits of CoRoT-7b at a lower activity level. *Astronomy & Astrophysics, 569*, A74.

Berdyugina, S. V. (1998). Surface imaging by the Occamian approach. Basic principles, simulations, and tests. *Astronomy & Astrophysics, 338*, 97.

Berdyugina, S. V. (2005). Starspots: A key to the stellar dynamo. *Living Reviews in Solar Physics, 2*, 8.

Bonomo, A. S., & Lanza, A. F. (2012). Starspot activity and rotation of the planet-hosting star Kepler-17. *Astronomy & Astrophysics, 547*, A37.

Borucki, W. J., Koch, D., Basri, G., Batalha, N., Brown, T., Caldwell, D., et al. (2010). Kepler planet-detection mission: Introduction and first results. *Science, 327*, 977.

Bryan, R. K., & Skilling, J. (1980). Deconvolution by maximum entropy, as illustrated by application to the jet of M87. *Monthly Notices of the Royal Astronomical Society, 191*, 69.

Chapman, G. A., Herzog, A. D., Lawrence, J. K., & Shelton, J. C. (1984). Solar luminosity fluctuations and active region photometry. *The Astrophysical Journal, 282*, L99.

Chapman, G. A., Cookson, A. M., Dobias, J. J., & Walton, S. R. (2001). An improved determination of the area ratio of faculae to sunspots. *The Astrophysical Journal, 555*, 462.

Chapman, G. A., Dobias, J. J., & Arias, T. (2011). Facular and sunspot areas during solar cycles 22 and 23. *The Astrophysical Journal, 728*, 150.

Cody, A. M., Stauffer, J., Baglin, A., Micela, G., Rebull, L. M., Flaccomio, E., et al. (2014). CSI 2264: Simultaneous optical and infrared light curves of Young disk-bearing stars in NGC 2264 with CoRoT and Spitzer; evidence for multiple origins of variability. *Astronomical Journal, 147*, 82.

Collier Cameron, A. (1997). Eclipse mapping of late-type close binary stars. *Monthly Notices of the Royal Astronomical Society, 287*, 556.

Collier Cameron, A. (1992). Modelling Stellar photospheric spots using spectroscopy, in surface inhomogeneities on late-type stars. *Lecture Notes in Physics, 397*, 33.

Cowan, N. B., Fuentes, P. A., & Haggard, H. M. (2013). Light curves of stars and exoplanets: Estimating inclination, obliquity and albedo. *Monthly Notices of the Royal Astronomical Society, 434*, 2465.

Croll, B. (2006). Markov Chain Monte Carlo methods applied to photometric spot modeling. *Publications of the Astronomical Society of the Pacific, 118*, 1351.

Donahue, R. A., Dobson, A. K., & Baliunas, S. L. (1997a). Stellar active region evolution – I. Estimated lifetimes of chromospheric active regions and active region complexes. *Solar Physics, 171*, 191.

Donahue, R. A., Dobson, A. K., & Baliunas, S. L. (1997b). Stellar active region evolution – II. Identification and evolution of variance morphologies in CA II H+K time series. *Solar Physics, 171*, 211.

Dorren, J. D. (1987). A new formulation of the starspot model, and the consequences of starspot structure. *The Astrophysical Journal, 320*, 756.

Dumusque, X. (2014). Deriving stellar inclination of slow rotators using stellar activity. *The Astrophysical Journal, 796*, 133.

Dumusque, X., Boisse, I., & Santos, N. C. (2014). SOAP 2.0: A tool to estimate the photometric and radial velocity variations induced by stellar spots and plages. *The Astrophysical Journal, 796*, 132.

Eker, Z. (1994). Modeling light curves of spotted stars. *The Astrophysical Journal, 420*, 373.

Foukal, P. (1998). What determines the relative areas of spots and faculae on sun-like stars? *The Astrophysical Journal, 500*, 958.

Frasca, A., Fröhlich, H. -E., Bonanno, A., Catanzaro, G., Biazzo, K., & Molenda-Żakowicz, J. (2011). Magnetic activity and differential rotation in the very young star KIC 8429280. *Astronomy & Astrophysics, 532*, A81.

Fröhlich, H. -E. (2007). The differential rotation of ε Eri from MOST data. *Astronomische Nachrichten, 328*, 1037.

Fröhlich, H. -E., Küker, M., Hatzes, A. P., & Strassmeier, K. G. (2009). On the differential rotation of CoRoT-2a. *Astronomy & Astrophysics, 506*, 263.

Fröhlich, H. -E., Frasca, A., Catanzaro, G., Bonanno, A., Corsaro, E., Molenda-Żakowicz, J., et al. (2012). Magnetic activity and differential rotation in the young Sun-like stars KIC 7985370 and KIC 7765135. *Astronomy & Astrophysics, 543*, A146.

García-Alvarez, D., Lanza, A. F., Messina, S., Drake, J. J., van Wyk, F., Shobbrook, R. R., et al. (2011). Starspots on the fastest rotators in the β Pictoris moving group. *Astronomy & Astrophysics, 533*, A30.

Gondoin, P. (2008). Contribution of Sun-like faculae to the light-curve modulation of young active dwarfs. *Astronomy & Astrophysics, 478*, 883.

Gray, D. F. (2008). In D. F. Gray (Ed.), *The observation and analysis of stellar photospheres*. Cambridge: Cambridge University Press.

Harmon, R. O., & Crews, L. J. (2000). Imaging stellar surfaces via matrix light-curve inversion. *The Astronomical Journal, 120*, 3274.

Haywood, R. D., Collier Cameron, A., Queloz, D., Barros, S. C. C., Deleuil, M., Fares, R., et al. (2014). Planets and stellar activity: Hide and seek in the CoRoT-7 system. *Monthly Notices of the Royal Astronomical Society, 443*, 2517.

Herrero, E., Lanza, A. F., Ribas, I., Jordi, C., & Morales, J. C. (2013). Photospheric activity, rotation, and magnetic interaction in LHS 6343 A. *Astronomy & Astrophysics, 553*, A66.

Herrero, E., Lanza, A. F., Ribas, I., Jordi, C., Collier Cameron, A., & Morales, J. C. (2014). Doppler-beaming in the Kepler light curve of LHS 6343 A. *Astronomy & Astrophysics, 563*, A104.

Huber, K. F., Czesla, S., Wolter, U., & Schmitt, J. H. M. M. (2010). Planetary eclipse mapping of CoRoT-2a. Evolution, differential rotation, and spot migration. *Astronomy & Astrophysics, 514*, A39.

Jetsu, L. (1996). The active longitudes of λ Andromedae, σ Geminorum, II Pegasi and V 711 Tauri. *Astronomy & Astrophysics, 314*, 153.

Kallrath, J., & Milone, E. F. (1999). In J. Kallrath, E. F. Milone (Eds.), *Eclipsing binary stars: Modeling and analysis* (Astronomy and Astrophysics Library). New York: Springer.

Kolláth, Z., & Oláh, K. (2009). Multiple and changing cycles of active stars. I. Methods of analysis and application to the solar cycles. *Astronomy & Astrophysics, 501*, 695.

Kovari, Z., & Bartus, J. (1997). Testing the stability and reliability of starspot modelling. *Astronomy & Astrophysics, 323*, 801.

Kővári, Z., & Oláh, K. (2014). Observing dynamos in cool stars. *Space Science Reviews, 186*, 457.

Lanza, A. F. (2015). Star-planet interactions. In *18th Cambridge Workshop on Cool Stars, Stellar Systems, and the Sun* (p. 811). http://www2.lowell.edu/workshops/coolstars18/articles/100-Lanza_CS18.pdf.

Lanza, A. F., Catalano, S., Cutispoto, G., Pagano, I., & Rodono, M. (1998). Long-term starspot evolution, activity cycle and orbital period variation of AR Lacertae. *Astronomy & Astrophysics, 332*, 541.

Lanza, A. F., Rodonò, M., Pagano, I., Barge, P., & Llebaria, A. (2003). Modelling the rotational modulation of the Sun as a star. *Astronomy & Astrophysics, 403*, 1135.

Lanza, A. F., Rodonò, M., & Pagano, I. (2004). Multiband modelling of the Sun as a variable star from VIRGO/SoHO data. *Astronomy & Astrophysics, 425*, 707.

Lanza, A. F., Piluso, N., Rodonò, M., Messina, S., & Cutispoto, G. (2006). Long-term starspot evolution, activity cycle, and orbital period variation of V711 Tauri (HR 1099). *Astronomy & Astrophysics, 455*, 595.

Lanza, A. F., Bonomo, A. S., & Rodonò, M. (2007). Comparing different approaches to model the rotational modulation of the Sun as a star. *Astronomy & Astrophysics, 464*, 741.

Lanza, A. F., Pagano, I., Leto, G., Messina, S., Aigrain, S., Alonso, R., et al. (2009a). Magnetic activity in the photosphere of CoRoT-Exo-2a. Active longitudes and short-term spot cycle in a young Sun-like star. *Astronomy & Astrophysics, 493*, 193.

Lanza, A. F., Aigrain, S., Messina, S., Leto, G., Pagano, I., Auvergne, M., et al. (2009b). Photospheric activity and rotation of the planet-hosting star CoRoT-4a. *Astronomy & Astrophysics, 506*, 255.

Lanza, A. F., Bonomo, A. S., Moutou, C., Pagano, I., Messina, S., Leto, G., et al. (2010). Photospheric activity, rotation, and radial velocity variations of the planet-hosting star CoRoT-7. *Astronomy & Astrophysics, 520*, A53.

Lanza, A. F., Bonomo, A. S., Pagano, I., Leto, G., Messina, S., Cutispoto, G., et al. (2011). Photospheric activity, rotation, and star-planet interaction of the planet-hosting star CoRoT-6. *Astronomy & Astrophysics, 525*, A14.

Lanza, A. F., Das Chagas, M. L., & De Medeiros, J. R. (2014). Measuring stellar differential rotation with high-precision space-borne photometry. *Astronomy & Astrophysics, 564*, A50.

Lehtinen, J., Jetsu, L., Hackman, T., Kajatkari, P., & Henry, G. W. (2011). The continuous period search method and its application to the young solar analogue HD 116956. *Astronomy & Astrophysics, 527*, A136.

Lindborg, M., Mantere, M. J., Olspert, N., Pelt, J., Hackman, T., Henry, G. W., et al. (2013). Multiperiodicity, modulations and flip-flops in variable star light curves. II. Analysis of II Pegasus photometry during 1979–2010. *Astronomy & Astrophysics, 559*, A97.

Lockwood, G. W., Skiff, B. A., Henry, G. W., Henry, S., Radick, R. R., Baliunas, S. L., et al. (2007). Patterns of photometric and chromospheric variation among sun-like stars: A 20 year perspective. *Astrophysical Journal Supplement Series, 171*, 260.

McQuillan, A., Aigrain, S., & Mazeh, T. (2013). Measuring the rotation period distribution of field M dwarfs with Kepler. *Monthly Notices of the Royal Astronomical Society, 432*, 1203.

McQuillan, A., Mazeh, T., & Aigrain, S. (2014). Rotation periods of 34,030 Kepler main-sequence stars: The full autocorrelation sample. *Astrophysical Journal Supplement Series, 211*, 24.

Messina, S. (2008). Long-term magnetic activity in close binary systems. I. Patterns of color variations. *Astronomy & Astrophysics, 480*, 495.

Mosser, B., Baudin, F., Lanza, A. F., Hulot, J. C., Catala, C., Baglin, A., et al. (2009). Short-lived spots in solar-like stars as observed by CoRoT. *Astronomy & Astrophysics, 506*, 245.

Narayan, R., & Nityananda, R. (1986). Maximum entropy image restoration in astronomy. *Annual Review of Astronomy and Astrophysics, 24,* 127.

Nutzman, P. A., Fabrycky, D. C., & Fortney, J. J. (2011). Using star spots to measure the spin-orbit alignment of transiting planets. *The Astrophysical Journal, 740,* L10.

Oshagh, M., Santos, N. C., Boisse, I., Boué, G., Montalto, M., Dumusque, X., et al. (2013). Effect of stellar spots on high-precision transit light-curve. *Astronomy & Astrophysics, 556,* A19.

Piskunov, N. E., Tuominen, I., & Vilhu, O. (1990). Surface imaging of late-type stars. *Astronomy & Astrophysics, 230,* 363.

Poe, C. H., & Eaton, J. A. (1985). Starspot areas and temperatures in nine binary systems with late-type components. *The Astrophysical Journal, 289,* 644.

Press, W. H., Teukolsky, S. A., Vetterling, W. T., & Flannery, B. P. (2002). *Numerical recipes in C++: The art of scientific computing* by William H. Press. xxviii, 1,002 p. : ill. ; 26 cm. Includes Bibliographical References and Index. ISBN:0521750334.

Rauer, H., Catala, C., Aerts, C., Appourchaux, T., Benz, W., Brandeker, A., et al. (2014). The PLATO 2.0 mission. *Experimental Astronomy, 38,* 249.

Reinhold, T., Reiners, A., & Basri, G. (2013). Rotation and differential rotation of active Kepler stars. *Astronomy & Astrophysics, 560,* A4.

Rice, J. B., & Strassmeier, K. G. (2000). Doppler imaging from artificial data. Testing the temperature inversion from spectral-line profiles. *Astronomy & Astrophysics Supplement Series, 147,* 151.

Roettenbacher, R. M., Monnier, J. D., Harmon, R. O., Barclay, T., & Still, M. (2013). Imaging starspot evolution on Kepler target KIC 5110407 using light-curve inversion. *The Astrophysical Journal, 767,* 60.

Rodono, M., Cutispoto, G., Pazzani, V., Catalano, S., Byrne, P. B., Doyle, J. G., et al. (1986). Rotational modulation and flares on RS CVn and BY Dra-type stars. I - Photometry and SPOT models for BY Dra, AU Mic, AR Lac, II Peg and V 711 Tau (= HR 1099). *Astronomy & Astrophysics, 165,* 135.

Rodonò, M., Cutispoto, G., Lanza, A. F., & Messina, S. (2001). The Catania automatic photo-electric telescope on Mt. Etna: a systematic study of magnetically active stars. *Astronomische Nachrichten, 322,* 333.

Sanchis-Ojeda, R., & Winn, J. N. (2011). Starspots, spin-orbit misalignment, and active latitudes in the HAT-P-11 exoplanetary system. *The Astrophysical Journal, 743,* 61.

Sanchis-Ojeda, R., Winn, J. N., & Fabrycky, D. C. (2013). Starspots and spin-orbit alignment for Kepler cool host stars. *Astronomische Nachrichten, 334,* 180.

Savanov, I. S., & Strassmeier, K. G. (2005). Surface imaging with atomic and molecular features. I. A new inversion technique and first numerical tests. *Astronomy & Astrophysics, 444,* 931.

Savanov, I. S., & Strassmeier, K. G. (2008). Light-curve inversions with truncated least-squares principal components: Tests and application to HD 291095 = V1355 Orionis. *Astronomische Nachrichten, 329,* 364.

Schneider, J. (2000). *The scientific potential of high precision transits of giant planets.* From Giant Planets to Cool Stars, ASP Conference Series (Vol. 212, p. 284).

Silva, A. V. R. (2003). Method for spot detection on solar-like stars. *The Astrophysical Journal, 585,* L147.

Silva-Valio, A. (2008). Estimating stellar rotation from starspot detection during planetary transits. *The Astrophysical Journal, 683,* L179.

Silva-Valio, A., Lanza, A. F., Alonso, R., & Barge, P. (2010). Properties of starspots on CoRoT-2. *Astronomy & Astrophysics, 510,* A25.

Silva-Valio, A., & Lanza, A. F. (2011). Time evolution and rotation of starspots on CoRoT-2 from the modelling of transit photometry. *Astronomy & Astrophysics, 529,* A36.

Strassmeier, K. G. (2009). Starspots. *Astronomy and Astrophysics Review, 17,* 251.

Strassmeier, K. G., & Bopp, B. W. (1992). Time-series photometric SPOT modeling. I - Parameter study and application to HD 17433 = VY ARIETIS. *Astronomy & Astrophysics, 259,* 183.

Strassmeier, K. G., & Olah, K. (1992). On the starspot temperature of HD 12545. *Astronomy & Astrophysics, 259,* 595.

Strassmeier, K. G., Bartus, J., Cutispoto, G., & Rodono, M. (1997). Starspot photometry with robotic telescopes: Continuous UBV and V(RI)_C photometry of 23 stars in 1991–1996. *Astronomy & Astrophysics Supplement Series, 125*, 11.

Tikhonov, A. N., & Goncharsky, A. V. (1987). *Ill-posed problems in the natural sciences* (344 p.). Advances in Science and Technology in the USSR. Mathematics and Mechanics Series. Moscow: MIR Publishers.

Titterington, D. M. (1985). General structure of regularization procedures in image reconstruction. *Astronomy & Astrophysics, 144*, 381.

Vogt, S. S., Penrod, G. D., & Hatzes, A. P. (1987). Doppler images of rotating stars using maximum entropy image reconstruction. *The Astrophysical Journal, 321*, 496.

Wolter, U., Schmitt, J. H. M. M., Huber, K. F., Czesla, S., Müller, H. M., Guenther, E. W., et al. (2009). Transit mapping of a starspot on CoRoT-2. Probing a stellar surface with planetary transits. *Astronomy & Astrophysics, 504*, 561.

Chapter 4
Reconstruction of Thermal and Magnetic Field Structure of the Solar Subsurface Through Helioseismology

K.M. Hiremath

Abstract Before the era of helioseismology, thermal and magnetic field structure of the Sun's subsurface was understood from the stellar structure evolutionary calculations and modeling of magnetic field structure. Recent discovery of Sun's oscillations whose amplitudes and frequencies are estimated with an unprecedented precision heralded a new era in solar physics that unraveled the hitherto unknown mysteries of the Sun's thermal and magnetic field structure of the subsurface. With a brief introduction to surface observations, I present the thermal structure of the solar subsurface as obtained from the evolutionary calculations. Both the thermal and magnetic field structures of the solar subsurface as reconstructed from helioseismology are also presented. Finally, I briefly dwell upon subtleties of different models that are developed for understanding both the thermal and magnetic field structures of the solar surface and subsurface respectively.

4.1 Introduction

Study of the Sun is important for very simple reason that Sun is very close to us so that details of its surface structure can be resolved to understand its internal structure. When I say Sun's internal structure means, shape of the Sun is nearly balanced by the three major forces, *viz.,* gravity, thermal, dynamic and magnetic forces, in decreasing order in the Sun's interior. Presently Sun's shape is dictated by the gravity and the thermal pressure that is generated by the nuclear energy generation, although in the early history of solar system formation, magnetic field structure might have also played a dominant role. Imbalance of either of these forces completely destroys the present shape and structure of the Sun.

Another reason to give importance for the Sun is, its whole or part of its structure can be treated as a gigantic laboratory wherein some of the known physics that can not be tested from the laboratory can be tested in the Sun. For understanding

K.M. Hiremath (✉)
Indian Institute of Astrophysics, Bengaluru-560034, Karnataka, India
e-mail: hiremath@iiap.res.in

© Springer International Publishing Switzerland 2016
J.-P. Rozelot, C. Neiner (eds.), *Cartography of the Sun and the Stars*,
Lecture Notes in Physics 914, DOI 10.1007/978-3-319-24151-7_4

Sun's structure, a knowledge of micro and macro physics are required. Some of the hypothetical physical phenomena that can not be tested in the laboratory can be tested from the astrophysical gigantic laboratory-the Sun. It is instructive to know that, for the first time, general relativistic phenomenon, such as bending of light near the massive body, is tested from the solar eclipse observations. Consensus emerged from the recent studies of Sun's surface dynamics and oscillations that, precession of the Mercury's orbit could be due to combined effect of a fast rotating core (nearly 2–3 times surface rotation; see Woodard 1984; Garcia et al. 2007), strong ($\sim 10^5$ G) magnetic field structure (Sturrock and Gilvarry 1967; Paterno et al. 1996), planetary gravitational attractions (Xu et al. 2011) and general relativistic effects.

Microphysics such as particle physics, equation of state, opacity, nuclear reactions are tested by considering the Sun as a laboratory. Neutrino, hitherto considered to be a massless particle from the standard models, has a small mass, that is concluded from the expected dearths of solar neutrinos as measured by the solar neutrino experiments. Recently Sun's interior is considered as a test bed for detection of the WIMPS (Weakly interacting massive particles), dark matter candidates (Turck-Chièze and Lopes 2012; Lopes et al. 2014; Lopes and Silk 2014; Turck-Chièze et al. 2012; Casanellas and Lopes 2014; Vincent et al. 2015) and possibly cosmic gravitational waves (Lopes and Silk 2014; Siegel and Markus 2014; McKernan et al. 2015) whose presence in the solar interior, if confirmed, may resolve some of the cosmological problems.

We are close to a nearest star such that influence in terms of radiation and high energy particles emitted by the Sun influence our environments and affect the life time of the artificial satellites that are used for human welfare and advancement. Space weather, disturbances in the geospace, such as ionosphere and magnetosphere of the Earth, is solely due to Sun only. Recent overwhelming evidences from the present and paleoclimatic data show that there is a strong solar forcing on the Earth's climate, such as temperature and Monsoon rainfall (Hiremath and Mandi 2004; Hiremath 2006a,b, 2009; Hiremath et al. 2015) of the Indian subcontinent, East Asian and Australia.

Hence, study of Sun is not only important for the intellectual and philosophical reasons, but also for humans' survival on the Earth itself. With this brief introduction to importance to the Sun, in Sect. 4.2, I elaborate in detail the observed solar cycle and activity phenomena. Section 4.3 introduces thermal structure of the solar interior as understood by solar structure evolutionary calculations. Whereas, with a brief introduction to helioseismology, Sect. 4.4 is reserved for reconstruction of thermal and magnetic field structures of the solar subsurface from helioseismic surface observations. Concluding remarks are presented in the last section.

4.2 Solar Cycle and Activity Phenomena

Before understanding the solar cycle and activity phenomena, let us know the magnitudes of standard estimates of different physical parameters such as mass, radius, etc., of the Sun. According to astronomical description, Sun is a G type main sequence star (spectral type of G2V), with a visual (apparent) magnitude of -26.74 and absolute magnitude (a hypothetical apparent magnitude if Sun's brightness is measured from 32.6 light years away form the Sun) of 4.83. Estimated distance between the Sun and the Earth (or 1 Astronomical Unit) is 1.496×10^8 km. Sun's mass and radius are estimated to be $\sim 2 \times 10^{30}$ kg and 7×10^5 km respectively. With effective temperature of ~ 6000 K, average density of Sun is found to be ~ 1408 kg/m^3. Total amount of energy emitted or luminosity of the Sun is 3.85×10^{26} J/s.

4.2.1 Sunspots

The most outstanding activity of the Sun is Sunspots (Fig. 4.1) that are dark and are cooler compared to the ambient medium. First observations of these blemishes of the Sun through telescopes are done independently by four Europeans, *viz.*, Galileo Galilee, Thomas Harriet, Christopher Schiener and Johannes Fabricius. Right from Galileo's observations, genesis of sunspots and their periodic behavior on the Sun is still a mystery and one of the unsolved problem in astrophysics.

4.2.2 Sunspot 11 Year Cycle

In 1843, Schwabe discovered that, on the Sun's surface, occurrence number of sunspots with different years is not constant and has a near periodicity of 11 years (Fig. 4.2, lower panel). After the 20 years of discovery of 11 year sunspot cycle, Richard Carrington's observations indicate that, on the surface of the Sun, occurrence position of sunspots is not random. As sunspot cycle progresses, during

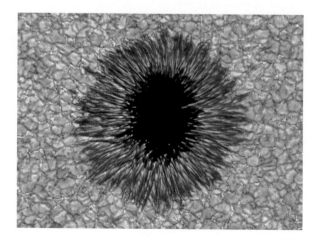

Fig. 4.1 Sunspot, courtesy BBSO/New Jersey Institute of Technology's New Solar Telescope

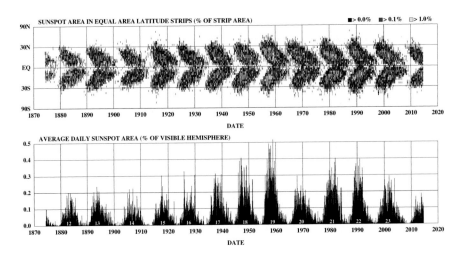

Fig. 4.2 For different years, sunspot occurrence area latitude-time variation represented as a butterfly diagram (*upper panel*) and daily variation of area (*lower panel*). Courtesy: David Hathaway

minimum activity periods, sunspots occur at high latitudes (particularly confined approximately to 50° north and south of the equator) and progress (drift) towards the solar equator. This spatial and temporal sunspot occurrence activity is represented as well known butterfly diagram (Maunder 1904) and is presented in Fig. 4.2 (upper panel). From long stretch of sunspot data, the most startling discovery by Eddy (1976) is that, during the period of 1645–1715, there was a dearth of occurrence of sunspot activity. Now such a dearth of sunspot activity phenomenon is popularly known as Maunder minimum of solar activity. From the well recorded sunspots data, such episodes of minimum activities, with different magnitudes are also found during different cycles of solar activity. Interestingly, Eddy (1976) also found that such an episode of dearth of sunspots activity was coincided with the *little ice age* when different parts of the European countries experienced sever cold weather.

4.2.3 Sunspot 22 Year Magnetic Cycle

From his ingenious experiment and due to Zeeman effect, Hale (1908) discovered that sunspots are associated with a strong magnetic field ($\sim 10^3$ G) structure. Recent analysis of MDI magnetograms obtained from space suggests that majority of the sunspots are bipolar (i.e., negative and positive polarities; Hiremath and Lovely 2007) during a particular sunspot cycle. Leading sunspots in the northern hemisphere have one polarity (say positive) and in the southern hemisphere the same have an opposite polarity (say negative). During the next sunspot cycle, in a particular hemisphere, polarity of the previous cycle reverses and returning back

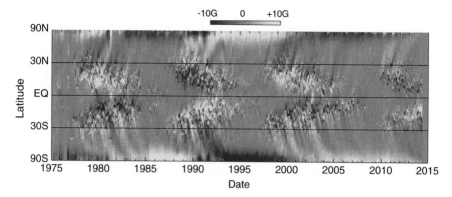

Fig. 4.3 A magnetic butterfly diagram: Latitude-time variation of longitudinally averaged radial magnetic field obtained from the Kitt Peak and SOHO data (Courtesy: David Hathaway)

to same polarity in the next cycle, thus constituting a 22 year magnetic cycle . For nearly four solar cycles (1975–2015), Sun's such a magnetic activity obtained from the Kitt Peak and SOHO magnetograms data is represented in Fig. 4.3. From this figure, one can notice the Hale's Polarity Law, polar field reversals, and the migration of higher latitude magnetic field structures toward the poles.

4.2.4 Spherical Harmonic Fourier Analysis of Sun's Magnetic Activity

In addition to a strong localized sunspots' magnetic field structure, Sun also possesses a weak ($\sim$1 G) (Hale 1908, 1913; Hale et al. 1918; Babcock 1947, 1961) large-scale general magnetic field structure that can clearly be discerned at the time of total solar eclipse and, during minimum solar activity period when one can notice intensity rays emanating from both the poles of the Sun that appear to be a dipole like structure. With his invented magnetograph, Babcock (1953) initiated mapping of global magnetic field structure of the Sun that senses both the background weak and the localized strong magnetic field (see Fig. 4.3). Now a days, important ground and space based observatories routinely take such magnetograms.

Spherical harmonic Fourier (SHF) analysis (Stenflo and Vogel 1986; Stenflo 1988; Knaack and Stenflo 2005) of magnetograms and magnetic field inferred from the sunspots (Gokhale et al. 1990; Gokhale and Javaraiah 1992) show that the axisymmetric global odd degree modes with selected periods ($\sim$22 yr and smaller) contribute predominantly to the evolution of the large-scale photospheric magnetic field structure of the Sun. The power spectra of these data show that the odd and even degree modes behave differently. All the odd parity modes have same periodicity of 22 years and the frequency of even parity modes increases with degree l that is almost similar to the observed helioseismic p mode $l - \nu$ diagnostic spectrum (see

Fig. 4.6). Recent SHF analysis (DeRosa et al. 2012) of many years of magnetograms suggests that, compared to magnetic energy of the toroidal (due to sunspots) flux, magnetic energy due to poloidal field is substantial and dominant for maintaining the large-scale solar cycle and activity phenomena.

4.2.5   Polar Faculae and Coronal Holes

After the well known sunspots, other outstanding solar cycle and activity phenomena, especially that occur near Sun's both the poles are *faculae* at the photosphere, extension of the same at the chromospheric heights as *plages* and much bigger activity phenomenon as *coronal holes*. It is interesting to note that all the three activity phenomena, *viz.,* polar faculae, coronal holes and the sunspots have same cyclic near periodicity of 11 years (Ikhsanov and Tavastsherna 2013; Karna et al. 2014). Polar faculae and the coronal holes have the same phase occurrences (Mordinov and Yazov 2014). Compared to sunspot occurrence activity, both the occurrence activity of polar faculae and the coronal holes, differ by phase difference of 5–6 years and appear in advance near the poles. In addition, both the polar faculae and the coronal holes follow the same phase as that of occurrence of polar magnetic flux (Karna et al. 2014). Both the Figs. 4.4 and 4.5 illustrate the low latitude (sunspots) and high latitude (polar faculae and polar coronal holes) activities respectively.

Therefore any physical magnetic model must be consistent to address the following important problems: (1) a large-scale weak ($\sim$1 G) dipole like magnetic field structure, (2) why Sun oscillates with a periodicity of either 11 year sunspot occurrence number or 22 year magnetic cyclic activity, (3) sunspot butterfly diagram, (4) reversal of the sun's magnetic field structure, (5) excitation of odd ($\sim$22 years) and even degree ($\sim$2–5 years) magnetic modes and, (6) how to explain

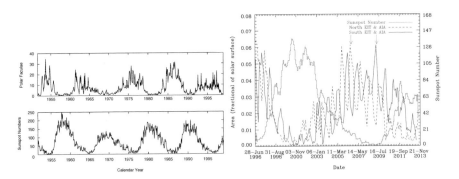

Fig. 4.4 *Left figure*: Yearly variation of polar faculae (*upper panel*) and the sunspot numbers (Courtesy: Linhua Deng). *Right figure*: Yearly variation of sunspot number (*black continuous line*), polar coronal hole activity (*blue and red colors*) (Courtesy: Nishu Karna)

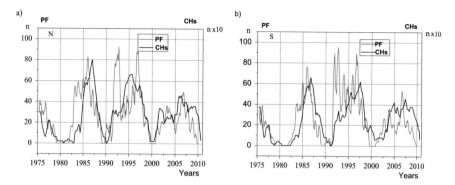

Fig. 4.5 Yearly variation of polar faculae (PF) and coronal holes (CHs) for the northern (*left figure*) and southern (*right figure*) hemispheres respectively

the occurrence of polar activity ($\sim$11 year sunspot or 22 year magnetic periodicity) that is antiphase with the equatorial sunspot activity.

4.3   Solar Thermal Structure as Understood from the Evolutionary Calculations

Vogt-Russell theorem states that star's internal structure can be uniquely determined by its mass and internal chemical composition if star is in hydrostatic and thermal equilibrium. With the assumption that Sun is in hydrostatic and thermal equilibrium, macro and micro physics are used to obtain the internal structure of the Sun. As for macrophysics, conservation of mass, hydrostatic equilibrium, depending upon physical condition of the solar interior, *viz.,* radiative or convective envelope, equation of energy transport and equation of thermal equilibrium (that means luminosity or total energy radiated by the Sun should be balanced by the energy generated by nuclear reactions at the core) are used. Mathematically all these macrophysical equations are compiled as follows

$$\frac{dM}{dr} = 4\pi r^2 \rho \,, \tag{4.1}$$

$$\frac{dP}{dr} = -\frac{GM\rho}{r^2} \,, \tag{4.2}$$

$$\frac{dL}{dr} = 4\pi r^2 \rho \epsilon \,, \tag{4.3}$$

$$\frac{dT}{dr} = -\frac{3}{4ac}\frac{\kappa\rho}{T^3}\frac{L}{4\pi r^2} \quad \text{if radiative},$$

$$= (\frac{dT}{dr})_{conv} \quad \text{if convective}, \tag{4.4}$$

$$\frac{dY}{dt} = \frac{dX}{dt} + \epsilon, \tag{4.5}$$

where the radial variables M, P, L, T, X and Y are the mass, pressure, luminosity, temperature, hydrogen abundance and helium abundance respectively. Other variables ϵ, κ and $(\frac{dT}{dr})_{conv}$ are the rate of nuclear energy generation, opacity of matter and knowledge of convective energy transport in the outer 30 % radius of the Sun. Further auxiliary equations are invoked from the microphysics such as equation of state, equations of opacity (that impedes flow of radiative energy) and rate of nuclear energy generation respectively. One can notice from Eq. (4.5) that changes in stellar structure basically depends upon the changes in hydrogen abundance into higher elements due to nuclear fusion reactions. In order to simplify the solutions, reasonable assumptions and approximations are made.

While evolving the afore mentioned equations, importance of Sun's rotational and magnetic field structure are neglected. Mass loss and accretion during the early history of Sun's evolution is completely neglected. In the first 70 % of solar radius from the center, energy generated from the deep core by nuclear reactions is transferred by radiation transport mechanism and in the next 30 % of the solar radius, energy is transferred by convection. In fact, it is also evident from the surface observations that Sun is fully convective at the outer 30 % of the Sun's radius. Hence, in the convective envelope, energy is transferred by the convection and is reradiated near the surface. Transfer of radiation by convection is treated by a mixing length parameter α (a ratio of the mixing length to the local pressure scale height; Böhm-Vitense 1958) which is considered to be a free parameter. Except helium, most of the Sun's chemical abundances (including lighter element Hydrogen and heavier elements or metals according to astronomers) are available from the spectroscopic inferences.

Hence, in addition to the mixing length parameter α, Helium abundance is also used as a free parameter in the evolutionary calculations. By adjusting the parameter α and the helium abundance, all the above equations are evolved up to the present age and the presently observed luminosity and radius are matched. Owing to energy generation due to nuclear fusion of hydrogen into helium, some part of the energy generated is also carried away to the space by the neutrinos that can indirectly be measured at the Earth.

With such a solar standard model, thermal structure (pressure, temperature, density, etc.,) of the Sun's interior is presented in Fig. 4.8 (blue continuous line).

The evolutionary computations of the standard solar model yield the following important salient physical structure of the Sun's interior. (1) Presence of dense, high temperature energy generating core. That means 50 % of solar mass and about 99 % of luminosity is concentrated with in 25 % of the solar radius from the center. High concentration of mass is attributed to the self gravitation and sharp radial gradient of the density. (2) Near the center, we have pressure of $\sim 10^{17}$ dynes/cm^2, temperature is $\sim 10^7$ K and density is ~ 150 gm/cm^3. (3) In the core, hydrogen abundance is depleted by 60 % and helium abundance is enhanced by 30 % compared to surface values. (4) From the center, moving towards surface, there is a sharp increase of opacity around $0.7 R_\odot$ (where $R_\odot$ is radius of the Sun) and increase of the radial gradient of temperature resulting in starting place for the convection. (5) Finally, standard solar model yields a sharp decrease in pressure near the surface. At the surface, the temperature is 6.4×10^3 K, the pressure is 0.14 atmospheres and the density is 3×10^{-7} gm/cm^3. (6) Observed neutrino flux emitted by the Sun and estimated at Earth is not compatible with the neutrino flux computed from this standard solar model, thus constitutes a solar neutrino problem.

4.4 Helioseismology

When there was no general consensus among the scientific community for understanding the following three unsolved solar physics problems (to name few), Helioseismic observations and inferences came in handy to resolve these outstanding issues. Three outstanding unsolved problems are: (1) physics of the solar sunspot butterfly diagram, (2) precession of the Mercury's orbit and, (3) the dearth of expected neutrinos emitted by the Sun. Before the era of helioseismology, in order that observed sunspots move from higher latitudes to the equator, kinematic turbulent dynamo models (Cowling 1953; Babcock 1961; Leighton 1964, 1969; Stix 1972; Krause 1976; Radler 1976; Yoshimura 1978; Duvall 2001; Ossendrijver 2003; Venkatakrishnan 2003; Dikpati 2005; Brandenburg and Subramaniyam 2005; Solanki et al. 2006; Hiremath and Lovely 2007 and references there in; Choudhuri 2008; Cally 2009; Hiremath 2010 and references there in; Charbonneau 2010; Sakurai 2012; Miesch 2012) require a rotation profile that increases from surface to the interior. Contrary to this expectation, helioseismic inferences rule out such a increasing rotation profile. As for Mercury's orbital precession, there is a difference of 43 s of arc per century which can not be accounted from the Newton's formalism of gravity. Hence, there were convincing proposals that Sun's core must be rotating fast (nearly 2–3 times surface rotation; see Woodard 1984; Garcia et al. 2007). Similarly unresolved issue of solar neutrino problem was argued possibly due to astrophysical solution. That means if Sun's core temperature is reduced, emission of neutrino flux is reduced. We will come to know in the subsequent section as to how helioseismic inferences of solar structure consistently resolved these conflicting issues.

Fig. 4.6 A typical l-v (l is
spherical harmonic degree
and v is frequency in milli
Hz) diagnostic diagram of the
p mode oscillations obtained
from the SOHO observations.
Different ridges slanted to x
axis are radial orders n and λ_h
is horizontal wavelength (in
Mega meter). Image courtesy
SOHO/MDI

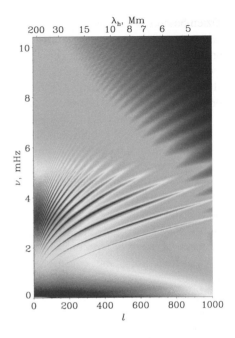

Observations from the Doplerograms show that Sun oscillates in millions (with
5 min as the dominant periodicity) of modes whose frequencies, phases and line
widths are measured accurately from the ground and space based experiments.
There are three main restoring forces, (viz., gravity, pressure and magnetic field
in decreasing order of magnitudes) that affect the physical state of the solar plasma.
The observed near 5 min oscillations are due to pressure (*p modes*) perturbations
of the solar interior. From the structure of the *degree (l)-frequency (v)* diagnostic
diagram (Deubner 1975) of the observed oscillations, it is unambiguously confirmed
(Ulrich 1970; Leibacher 1971) that these pressure perturbations are standing
oscillations in the solar interior. One such a *degree (l)-frequency (v)* diagnostic
diagram constructed from helioseismic observations is presented in Fig. 4.6 (see
also Chap. 2, page 26). Whereas higher periods around 160 min (Severnyi et al.
1976; Appourchaux and Palle 2013 and references there in) are probably due to
gravity and 1–22 year long period (Hiremath 1994 and references there in; Hiremath
and Gokhale 1995; Hiremath 2009, 2010 and references there in) oscillations could
be due to large-scale combined weak ($\sim$1 G) poloidal and a strong toroidal ($\sim$10–
10^3 G) magnetic field structure.

Depending upon physical conditions of the solar interior, all the three modes
are excited. *Sound waves,* due to pressure perturbations, are stochastically excited
by the turbulent region of the convective envelope (Gough 1985; Kumar et al.
1996; Belkacem et al. 2009). Other possible mechanisms for excitations of pressure
(*p*) modes are: (1) penetrative convection (Andersen 1996; Dintrans et al. 2005),
(2) mode coupling (Dziembowski 1983; Guenther and Demarque 1984; Ando
1986; Wentzel 1987; Wolff and O'Donovan 2007) and, (3) magnetic torques

(Dziembowski et al. 1985) respectively. Whereas *gravity* modes are excited in the stable radiative core. As the large-scale magnetic field structure is threaded throughout the solar interior, magnetic modes (poloidal and toroidal, depending upon whether large-scale structure is poloidal or toroidal magnetic field), due to Alfven wave perturbations, probably are excited throughout the Sun's interior (see for details Hiremath 2010 and references there in).

4.4.1 Inference of Thermal Structure by Comparing the Observed and Computed Frequencies

As 'p' modes are recognized to be standing oscillations in the solar interior, observed frequencies are matched with the theoretical frequencies that are obtained from linearizing the hydrodynamic equations (with the adiabatic constraint that excited 'p' modes do not exchange energy with the ambient medium). By using linearized hydrodynamic equations and with proper boundary conditions, standard solar model structure is used for computation of theoretical frequencies (Dalsgaard 2003; Unno et al. 1989) and are compared with the observed frequencies. It is found that, computed frequencies that are related to low and medium degree modes match very well with the observed frequencies. Whereas computed frequencies do not match very well with the observed high degree modes' frequencies that probably suggests that near surface physics is not very well understood.

4.4.2 Inference of Thermal Structure: Primary Inversions

In this method, standard solar model as a reference model, observed p mode frequencies are used to estimate the thermal structure, such as sound speed and density of the solar interior. As the observed amplitudes of 'p' mode oscillations are very small, it is reasonable to consider the adiabatic approximation for the solution of linearized hydrodynamic equations. With the constraint of conservation of mass, generalized form of equation (Lynden-Bell and Ostriker 1967; Gough and Taylor 1984; Unno et al. 1989; Dalsgaard 2003) of oscillation that takes into account the effects of velocity flows $\mathbf{v}$ and magnetic field structure $\mathbf{B}$ is given as follows

$$\mathcal{L}(\boldsymbol{\xi}) - \rho\omega^2\boldsymbol{\xi} + \nabla\delta p + \rho[\omega\mathcal{M}(\boldsymbol{\xi}) + \mathcal{N}(\boldsymbol{\xi}) + \mathcal{B}(\boldsymbol{\xi})] = 0, \qquad (4.6)$$

where

$$\mathcal{L}(\xi) = \nabla(c^2 \rho \nabla . \xi + \nabla P . \xi) - g\nabla.(\rho\xi) - G\rho\nabla(\int_V \frac{\nabla.(\rho\xi)dV}{|\mathbf{r} - \mathbf{r}'|}) , \qquad (4.7)$$

$$\mathcal{M}(\xi) = 2i[\boldsymbol{\Omega}_0 \times \xi + (\mathbf{v} \cdot \nabla)\xi] , \qquad (4.8)$$

$$\mathcal{N}(\xi) = (\boldsymbol{v} \cdot \nabla)^2 \xi - 2\boldsymbol{\Omega}_0 \times [(\xi \cdot \nabla)\mathbf{v} - (\mathbf{v} \cdot \nabla)\xi] - (\xi \cdot \nabla)(\mathbf{v} \cdot \nabla)\mathbf{v}, \qquad (4.9)$$

$$\mathcal{B}(\xi) = (4\pi\rho)^{-1}\big[[\rho^{-1}(\xi \cdot \nabla)\rho + \nabla \cdot \xi]\mathbf{B} \times (\nabla \times \mathbf{B})$$

$$- [(\nabla \times \mathbf{B}') \times \mathbf{B} + [(\nabla \times \mathbf{B}) \times \mathbf{B}']\big] , \qquad (4.10)$$

where $\mathcal{L}$ is differential operator, ξ is eigen function of oscillations, c is speed of sound, ρ is density, P is pressure, $\mathcal{M}(\xi)$ and $\mathcal{N}(\xi)$ are effects due to velocity flows, $\mathcal{B}(\xi)$ is effect due to magnetic field structure and other symbols have usual meanings. For the sake of simplicity, effects due to flows and magnetic field structure in the solar interior are neglected and, with special boundary conditions such that density and pressure perturbations completely vanish at the surface, Eq. (4.6) leads to the form $\mathcal{L}(\xi) = \rho\omega^2\xi$ which is an eigen value problem and is also Hermitian. By invoking the variational principle (Chandrasekhar 1964 see also Unno et al. 1989) this equation is linearized and the frequency difference $\delta\nu_{n,l}$ obtained from the Sun and the standard solar structure model (Basu 2010 and references there in) yields the following equation

$$\frac{\delta\nu_{n,l}}{\nu_{n,l}} = \int_0^R K^{nl}_{c^2,\rho}(r)\frac{\delta c^2}{c^2}(r)dr + \int_0^R K^{nl}_{\rho,c^2}(r)\frac{\delta\rho}{\rho}(r)dr + \frac{F(\nu_{n,l})}{E_{nl}} , \qquad (4.11)$$

where $K^{nl}_{c^2,\rho}(r)$ and $K^{nl}_{\rho,c^2}(r)$ are kernels that involve eigen functions of the oscillations and the known solar structure. In this equation, $\frac{\delta c^2}{c^2}$ and $\frac{\delta\rho}{\rho}$ are the normalized differences of sound speed and density between the Sun and the reference solar structure model. Whereas the last term (see for details Basu 2010) in the right hand side of above equation is a correction for the surface effects that may not be due to linearization of hydrodynamic equations. There are many inversion algorithms (Gough and Thompson 1991; Unno et al. 1989; Dalsgaard 2002 and references there in) to infer the sound speed and density. Typical inverted sound speed and density profiles from one such inversion technique (Basu et al. 2009) are presented in Fig. 4.7.

It is obvious from Fig. 4.7 that the inferred sound speed is almost similar to the sound speed computed from the standard solar model (in this case Dalsgaard's 1996 model) although there are statistically significant differences near base of the convection zone and in the radiative zone. The inferred sound speed (Fig. 4.7) is for the particular period of observation and can be represented as a steady part of thermal structure of the solar interior. However, recent overwhelming evidences show that observed p mode frequencies increase with the strength of solar activity, especially solar cycle effect is pronounced for the high frequency modes. If the

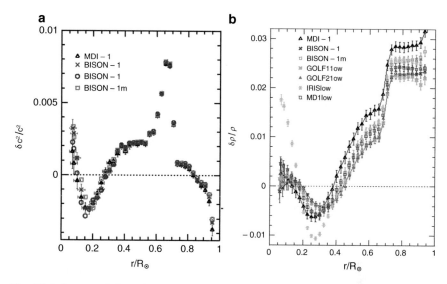

Fig. 4.7 Inference of thermal structure: (i) *Left figure* illustrates the sound speed difference between the Sun and the model and, (ii) *right figure* illustrates the density difference between the Sun and the model. Image courtesy, Basu

magnitudes of frequencies change from solar minimum to maximum, it is interesting to know whether such changes affect the inverted sound speed. In fact, Rabello-Soares (2012) inversion of the changing frequencies with the solar cycle indeed shows interesting statistically significant, oscillatory perturbations of normalized sound speed in the radial directions.

If one notice Fig. 4.7, information regarding thermal structure of the subsurface is not available after $0.95R_\odot$. However, from the f modes one can probe this near surface region. Infact, from the inversion of f mode frequencies, Lefebvre and Kosovichev (2005) and Lefebvre et al. (2007) found changes of thermal structure in this region with respect to solar cycle. Interestingly these authors found that for the subsurface region of 0.97 and $0.99R_\odot$, thermal structure varies in phase with the solar cycle, whereas above the region of $0.99R_\odot$, thermal structure varies in antiphase with the solar cycle. Infact such a transition layer near the surface is coined as *leptocline region* by Rozelot and Lefebvr (2006). Origin of such changes with the solar cycle is understood mainly due to changes in the sun's radius (Lefebvre et al. 2009).

4.4.3 Inference of Thermal Structure: Secondary Inversions

In Sect. 4.3, with appropriate boundary conditions and by evolving stellar structure equations, we obtained the internal structure (such as pressure, temperature, etc.,) of Sun. However, some of the so-called reasonable assumptions such as neglect of rotation and magnetic field can not be guaranteed during the early history of Sun's evolution. This is because, observations from the distant universe show that Sun like stars are fast rotating and have a very strong magnetic field structure whose effects can not be neglected in the structure equations. Similarly, prescribed mixing length parameter α and the helium abundance in the evolutionary calculations are too adhoc. In order to avoid these inconsistencies, is there any independent and self consistent way of getting the thermal structure of the Sun's interior, without any bias of Sun's evolutionary history? Yes, this question can be answered with affirmative way if one knows the sound speed of the solar interior. As sound speed is a thermodynamic variable which in turn depends upon pressure and temperature, by supplementing the knowledge of microphysics and by imposing the sound speed on the stellar structure equations, one can infer not only the chemical abundances (such as hydrogen X, helium Y and, heavier elements Z in principle) and other thermal structures such as pressure, temperature, etc., but also one can infer the mass, luminosity, depth of convection zone etc., from this method. Thus, such a model developed from the helioseismically inferred sound speed is called *"solar seismic model"* .

With appropriate boundary conditions and by imposing the helioseismically inverted sound speed (Takata and Shibahashi 1998), in the previous study (Shibahashi et al. 1995; Shibahashi and Takata 1996, 1997; Takata and Shibahashi 1998; Antia and Chitre 1999), stellar structure equations (4.1)–(4.4) are used to solve for the thermal structure of the radiative core. Although deduced internal structure profiles of seismic model are almost same as the structure profiles obtained from the standard solar model, there is no guarantee that luminosity and mass obtained from such a seismic model satisfy the observed surface mass and luminosity of the Sun.

Hence, with the same inverted sound speed (Takata and Shibahashi 1998), and in order to overcome this inconsistency, Shibahashi et al. (1998a,b, 1999) developed a seismic model that takes into account the whole (radiative core and the convective envelope) interior of the Sun. However, the deduced internal structure profiles, helium abundance and depth of the convection zone are not very different than the profiles obtained from the standard solar model.

Recently, we (Hiremath, Shibahashi and Takata) used helisoeismically imposed sound speed (kindly provided by Antia) and developed a seismic model whose structure profiles are presented in Fig. 4.8 (red dotted and dashed lines). One can notice from these profiles that thus developed seismic model is entirely not different than the solar standard model (in case Dalsgaard et al. 1996 standard solar model is considered). The perfect matching of these internal structure profiles could be due to same hydrogen to heavy element abundance ratio (X/Z) used in developing the seismic model. However, if we use the recently obtained (Asplund et al. 2004, 2009) and downgraded abundance ratio, we get the values of helium abundance and depth of the convection zone different than the values obtained from the conventional standard solar model and above mentioned solar seismic model. It is to be noted that seismic model can also be used for obtaining the solar neutrino fluxes. Hence, for the conventional $(X/Z = 0.0247)$ and lower $(X/Z = 0.0215, 0.0188, 0.0158)$ hydrogen

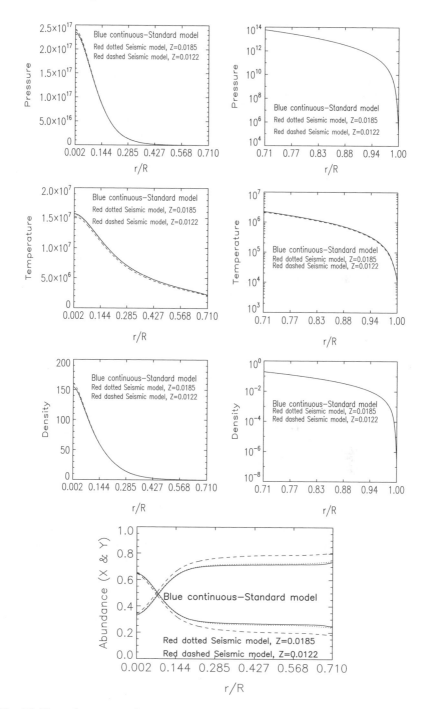

Fig. 4.8 Thermal structure such as pressure, temperature, density and hydrogen abundance X and helium abundance Y obtained by the solar seismic model for surface heavy elemental abundances $Z = 0.0185$ and $Z = 0.0122$ respectively. Units are in cgs scale

Table 4.1 Neutrino fluxes (in the units of $10^{10} \mathrm{cm}^{-2} \mathrm{s}^{-1}$) estimated by the solar seismic model

	SSM	Seism1	Seism2	Seism3	Seism4
	Z = 0.0185	Z = 0.0185	Z = 0.0165	Z = 0.015	Z = 0.0122
	Z/X = 0.0247	Z/X = 0.0247	Z/X = 0.0215	Z/X = 0.0188	Z/X = 0.0158
Source	BCZ = 0.709	BCZ = 0.709	BCZ = 0.709	BCZ = 0.709	BCZ = 0.72
pp	6.0	6.01	6.05	6.07	6.1
pep	0.014	0.015	0.014	0.015	0.015
hep	8×10^{-7}	0.3×10^{-9}	0.3×10^{-9}	1.3×10^{-7}	1.32×10^{-7}
^{7}Be	0.47	0.44	0.42	0.41	0.379
^{8}B	5.8×10^{-4}	4.5×10^{-4}	4.03×10^{-4}	3.76×10^{-4}	3.18×10^{-4}
^{13}N	0.06	0.057	0.047	0.04	0.0287
^{15}O	0.05	0.052	0.042	0.036	0.025
^{17}F	5.2×10^{-4}	4.02×10^{-4}	3.2×10^{-4}	2.7×10^{-4}	1.92×10^{-4}

and heavy element abundance ratio, in Table 4.1, we present deduced depth of the convection zone BCZ and the computed neutrino fluxes at the Earth. First to sixth columns in Table 4.1 are: (1) different nuclear reactions, (2) neutrino fluxes at the Earth estimated from standard model and, (3) rest of the columns represent seismic models that use different hydrogen to heavy elemental abundance ratio.

Before analyzing the results of neutrino fluxes for the low chemical abundance ratio, let us analyze the neutrino fluxes obtained from the seismic model (Seism1, third column) that uses the conventional chemical abundance ratio. One can notice from Table 4.1 that, for the earlier (Grevesse and Noels 1993) chemical abundance ratio $Z/X = 0.0247$, neutrino fluxes computed from the seismic model are almost same as the neutrino fluxes computed from the solar standard solar model. This is a main reason that deficiency of neutrinos emitted by the Sun lies in the solution of neutrino physics rather in the astrophysical solutions, *viz.*, changing of uncertain physics of the interior, such as opacity, equation of state, etc., or chemical composition (especially the heavy elemental abundance Z) that could not solve the solar neutrino problem.

Coming to the inclusion of low chemical abundance ratio for the development of the seismic model, computed neutrino fluxes, especially for ^{8}B, there is a substantial reduction and is almost similar to the observed neutrino fluxes. However, deduced depth of base of the convection zone is higher and the inferred helium abundance is very low ($\sim$0.18) which is inconsistent with the cosmic helium abundances ($\sim$0.23) and helium abundance as inferred from helioseismology (Kosovichev et al. 1992; Antia and Basu 1994; Basu and Antia 2004).

4.4.4 Inference of Magnetic Field Structure: Primary Inversions

As the Sun is rotating and is pervaded by a large-scale combined poloidal and toroidal magnetic field structure, magnitudes of frequency of the solar p modes are affected. Similar to Zeeman effect that splits the spectral lines due to a strong magnetic field structure, solar rotation splits the frequency of individual modes, especially odd degree modes. Hence, observed rotational frequency splittings are used to infer the internal rotation of the Sun. Details of inversion procedure can be found in the previous (Hiremath 2013 and references there in) review. The rotational profile of the solar interior as inferred by the helioseismic rotational splittings is presented in Fig. 4.9. Whereas even degree modes are influenced by a combined effect of rotation (a second order effect, first order rotational effect is on the odd degree modes), magnetic field and aspherical sound speed perturbations (Kuhn 1988) of the solar interior. Hence, by neglecting aspherical sound speed perturbations and by removing second order rotational effects in the even degree frequency splittings, resulting residual of even degree splittings are compared with the computed frequency splittings with a assumed form of magnetic field structure (Dziembowski and Goode 1989; Gough and Thompson 1991; Dziembowski and Goode 1991; Basu 1997; Antia et al. 2000a; Antia 2002; Baldner et al. 2010). With this method, Baldner et al. (2010) came to a conclusion that the observed

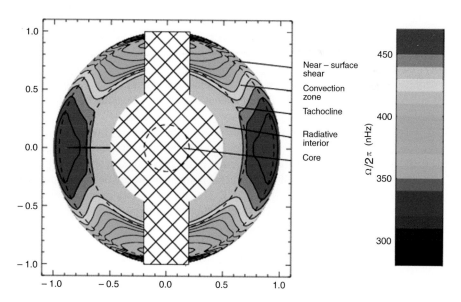

Fig. 4.9 Solar internal rotation profile as inferred from helioseismic 12 years MDI data. *Black thick continuous lines* are isorotational contours. One can notice two rotational gradients, near the surface and near base of the convection zone respectively. Whereas most of the radiative core rotates rigidly. (Courtesy: Howe)

variation of even degree frequency splittings for different modes can be explained if the Sun is pervaded by a right combination of poloidal ($\sim$100 G) and toroidal (that varies from $\sim 10^3$ G near the surface to 10^4 G near base of the convection zone) magnetic field structure. Interestingly, from the analysis of sunspot data during their initial appearance on the surface from the SOHO/MDI magnetograms, Hiremath and Lovely (2012) estimated strength of toroidal magnetic field structure of similar order confirming the previous theoretical estimates (Choudhuri and Gilman 1987; D'Silva and Choudhuri 1993; D'Silva and Howard 1994; Hiremath 2001; Brun et al. 2004).

4.4.5 Inversion of Poloidal Magnetic Field Structure: Secondary Inversions

Observations (Ambroz et al. 2009; Pasachoff 2009; Pasachoff et al. 2009) of white light pictures obtained during total solar eclipse, especially in the solar minimum activity, indicate that the delineating rays of intensity structures emanating from both the poles represent tracing of dipole like magnetic field structures. It is interesting to know whether Sun has retained such as large scale weak ($\sim$0.01–1 G) magnetic field structure probably of primordial origin. In fact, with the current limit of instrumental detection of solar observations, such a large-scale magnetic field of primordial origin can not be ruled out. From theoretical and observational (from the stars in pre-main sequence phase which are fast rotating and are magnetically active) point of views, it is reasonable to expect such a large scale magnetic field structure in the solar interior whose diffusion time scale is order of billions of years, larger than the Sun's age.

We have an advantage of helioseismic inferences of internal rotation rate of the plasma that geometry and magnitude of such a large-scale magnetic field structure in the solar interior can be inferred. Surface observations indicate that Sun's large-scale poloidal average magnetic field structure can not be greater than 1 G (Stenflo 1993). As the strength of large-scale poloidal magnetic field structure is very weak compared to magnitude of rotational structure of the Sun, one can show (Hiremath 1994; Hiremath and Gokhale 1995) that such a poloidal magnetic field structure can isorotate with the solar plasma. Hence, if Ω is angular velocity of the plasma and Φ is the magnetic flux function representing flux of the large-scale weak poloidal magnetic field structure, one can show that

$$\Omega = function(\Phi) . \tag{4.12}$$

To a first approximation, right hand side of equation can be linearized in the following way

$$\Omega = \Omega_0 + \Omega_1(\Phi) . \tag{4.13}$$

That means, if one knows the internal rotational profile of a star in general and Sun in particular, with a suitable combination of poloidal magnetic field structure and by subjecting to a least-square fit of the above equation, one can infer the Sun's large-scale magnetic field structure of the solar interior.

Previous studies (Hiremath 1994; Hiremath and Gokhale 1995) have shown that, for both the radiative core (RC) and the convective envelope (CE), poloidal component of the Sun's steady magnetic field structure can be modeled as an analytical solution of magnetic diffusion equation. Such a solution yields the following magnetic flux function (Φ) in RC

$$\Phi_{RC}(x, \vartheta) = 2\pi A_0 R_c^2 x^{1/2} sin^2\vartheta \sum_{n=0}^{\infty} \lambda_n J_{n+3/2}(\alpha_n x) C_n^{3/2}(\mu), \tag{4.14}$$

where $x = r/R_c$, R_c is radius of the radiative core, $\mu = cos\vartheta$, r and ϑ are radial and colatitude coordinates, n is non negative integer, $C_n^{3/2}$ is the Gegenbaur's polynomial of degree n, $J_{n+3/2}(\alpha_n x)$ is Bessel function of order $(n + 3/2)$ and α_n are the eigen values that are to be determined from the boundary conditions. Here A_0 is taken as a scale factor for the field and hence $\lambda_n = A_n/A_0$.

Similarly, in the region of convective envelope, magnetic flux function can be modeled as

$$\Phi_{CE}(x, \vartheta) = \pi B_0 R_c^2 sin^2\vartheta \left[x^2 C_0^{3/2}(\mu) + \sum_{n \geq 0} \hat{\mu}_{n+1} x^{-(n+1)} C_n^{3/2}(\mu) \right], \tag{4.15}$$

where $\hat{\mu}_{n+1} = M_n/(\pi B_0 R_c^{n+3})$ are strengths of multipoles that are scaled to an asymptotically uniform magnetic field structure B_0 as $x \to \infty$.

In case of the Sun, with the appropriate boundary conditions and available helioseismically inferred rotation rate (Dziembowski et al. 1989; Antia et al. 1998) the unknown coefficients and eigen values can be computed uniquely. With this information, geometry and radial variation of poloidal magnetic field structure (see Fig. 4.10) can also be reconstructed in the solar interior (Hiremath 1994; Hiremath and Gokhale 1995). Such a reconstructed poloidal component of steady part of magnetic field structure consists of a diffusing uniform like magnetic field structure in the radiative core and a combined dipole and quadrupole like magnetic field structure in the convective envelope.

It is interesting to know strength of poloidal magnetic field structure near the solar core. Physics of this region is essential to unravel some of mysteries posed by the observations. As multipole moments M_n that are used to reconstruct the poloidal magnetic field structure are scaled with the asymptotic uniform magnetic field structure (that merges with the interstellar magnetic field structure), if one accepts the magnitude of interplanetary magnetic filed structure ($B_0 \sim 10^{-5}$ G), strength of Sun's poloidal magnetic field structure near its center is estimated to be $\sim 10^7 - 10^8$ G. From the dynamical constraints point of view, this strong magnetic field structure near the center can not be accepted. However, if the 22 yr solar

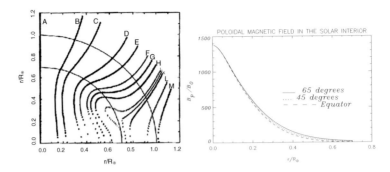

Fig. 4.10 *Left figure* illustrates the meridional cross section of poloidal component of magnetic field structure. This figure is reproduced from, and all rights are reserved by, the Astrophysical Journal. Where as *right figure* illustrates radial and latitudinal variations of the magnitude of the poloidal magnetic field structure B_p normalized to an asymptotic uniform magnetic field B_0 that merges with the interplanetary field structure

cycle and activity phenomena probably could be due to global slow MHD modes (Hiremath and Gokhale 1995), strength of asymptotically uniform magnetic field structure B_0 is found to be $\sim 10^{-2}$ G. With this magnitude, strength of poloidal magnetic field structure near the solar core is estimated to be $\sim 10^5$ G. Unfortunately, observed low degree solar modes do not show such a signature of strong magnetic field structure near the center. However, as low degree p modes sense the solar core region very poorly, it is not ruled out that such a strong magnetic field structure near the center might exists. This reality can be possible only when the g (due to buoyancy) modes are unambiguously detected from the surface observations or indirectly (for example characteristics of g modes or modified by the strong magnetic field; see Rashba et al. 2006, 2007; Rashba 2008; Burgess et al. 2004) with consistent way. Another promising way of detecting such a strong magnetic field structure is due to Sun's surface oblate measurements (Rozelot and Damiani 2011; Damiani et al. 2011; Rozelot et al. 2011; Rozelot and Fazel 2013; Meftah et al. 2015).

4.5 Concluding Remarks

Let us recapitulate the reconstruction of thermal and magnetic field structure of the solar subsurface through helioseismology. After giving a brief introduction to the solar cycle and activity phenomena, standard solar model and, helioseismology, thermal and magnetic field structure of the solar subsurface are reconstructed as deduced from helioseismology.

4.5.1 Recent Estimation of Chemical Abundances and Incompatible Seismic Models

Helioseismic inferences revolutionized our understanding of physics of the solar interior. As presented in this chapter, from helioseismic inferences, one of the parameter of thermal structure such as sound speed is obtained right from near surface up to close to the solar interior. From helisoeismically inferred sound speed, seismic model is developed in which we find that thermal structure of the solar interior is not entirely different than the structure computed from the standard solar model. However, thermal structure that is reconstructed form seismic solar models that incorporate the recent revision (Asplund et al. 2004, 2009) of hydrogen to heavy elemental abundance ratio is not compatible with the Sun's thermal structure. These results suggest that probably careful modeling and computation (Chaplin and Basu 2008; Basu and Antia 2008) in estimating the heavy elemental abundance are necessary. If more evidences lead to the general consensus about the heavy elemental abundance, there is a serious problem in explaining the sound speed difference between the models and the Sun near base of the convection zone and in the radiative interior. Probably, as deeply dwelt by Basu et al. (2014), in the recent review, physics (especially dynamics and magnetic field structure) of the radiative core has to be revisited and to be explored further.

4.5.2 Solar Cycle Changes of Frequencies and the Sound Speed

Recent intriguing result that frequencies (Woodard and Noyes 1985; Ronan et al. 1994; Howe et al. 1999; Antia et al. 2000b; Chaplin et al. 2001; Rabello-Soares et al. 2008; Jimenez et al. 2011; Tripathy et al. 2013) and hence the inverted sound speed (Rabello-Soares 2012) vary with the solar cycle. More inferences are accumulating that solar cycle variation of the frequencies due to high degree modes is solely due to variation of physics of the near surface. However, one has to explain the physics behind the solar cycle changes of the frequencies due to low degree modes.

4.5.3 Steady Parts of Poloidal and Toroidal Magnetic Field Structure

As for reconstruction of magnetic field structure of the solar interior from helioseismology, poloidal component of steady part of magnetic field structure is reconstructed. It is found that reconstructed poloidal magnetic field structure likely consists of a combined field structure that is dictated by diffusion (time scales are $\sim$ billion yrs) in the radiative core and a current free like structure that is dictated by the convection. From the constraint of 22 yrs magnetic cycle, strength of magnetic field structure near the solar core is estimated to be $\sim 10^5$ G. It is concluded that if signature of such a strong magnetic field structure is not detected from the low degree modes, one has to wait for unambiguous detection of g modes whose amplitudes are higher near the solar core.

This chapter also presents the estimates of magnitude ($\sim 10^4$ G near base of convection zone and ~ 100 G near the surface) of magnetic field, probably due to combined poloidal and toroidal magnetic field structure, from the p modes' even degree frequency splittings. Interestingly, near base of the convection zone, such a magnitude of magnetic field structure matches with theoretically estimated (Choudhuri and Gilman 1987; D'Silva and Choudhuri 1993; D'Silva and Howard 1994; Hiremath 2001; Brun et al. 2004; Hiremath and Lovely 2007) magnitude of toroidal magnetic field structure. As the well known sunspots are supposed to be originated from toroidal magnetic field structure of the convective envelope, effect due to such a field structure on the even degree frequency splittings has to be delineated. In addition, as many years of helioseismic observations are available, from the even degree mode splittings, steady and time dependent parts of toroidal magnetic field structure of the convective envelope are to be separated. Infact, further helioseismic inferences of time dependent part of toroidal magnetic field structure (due to which sunspots are supposed to be formed) of the whole convective envelope that varies from year to decadal scales is necessary to understand the origin of solar cycle and activity phenomena.

Another interesting topic to ponder over in this chapter is the reconstruction of steady part of poloidal magnetic field structure of the solar interior from the helioseismology. Unfortunately, as the strength of poloidal magnetic field structure is so weak (~ 1 G in the whole convective envelope; see Fig. 4.10) that signature of such a field structure can not be sensed from the 'p' modes. Hence, indirect information (either weak magnetic field structure isorotates with the solar internal rotation or from p modes even degree frequency splittings) is only available. That too in the present chapter, geometry and magnitude of steady component of poloidal magnetic field structure and, magnitude of probably a combined (poloidal and toroidal) magnetic field structure in the convective envelope is obtained. However, further helioseismic inferences on the time dependent part of poloidal part of magnetic field structure that has year to decadal time variations are necessary to understand the polar magnetic field activity that is antiphase with the middle latitude sunspot activity.

4.5.4 Origin and Consequences of Near Surface Rotation Profile and Toroidal Magnetic Field Structure

If future helioseismic inferences confirm the steady part of toroidal magnetic field structure in the solar interior, probably whose diffusion time scales is $\sim$ billion yrs, there will be important consequences for explanation of near surface shear rotation profile, accretion of the proto planetary mass on to the Sun and, faint young Sun paradox during early evolutionary history of the Sun.

For example, although many studies (Kitchatinov and Rüdiger 1995; Elliott et al. 2000; Robinson and Chan 2001; Hiremath 2001; Brun and Toomre 2002; Rempel 2005; Miesch et al. 2006; Kuker et al. 2011; Brun et al. 2011) reproduced the rotational isocontours as inferred by helioseismology, to the knowledge of this author, none of the studies reproduced isorotational counters close ($0.935R_\odot$ to $1.0R_\odot$) to the surface. Of course, we can not brush aside that this is a resolution problem in the numerical simulations. Even if future numerical simulations achieve this goal of reproducing near surface rotation profile, such a decreasing profile is unstable and does not satisfies the MHD stability criterion (Dubrulle and Knobloch 1993; Mestel 1999), $r^2 \frac{d\Omega^2}{dr} - (\frac{1}{r^2}) \frac{d}{dr}[\frac{(r^2 B_\phi^2)}{4\pi\rho}] > 0$ (where r is the radial coordinate, Ω is angular velocity, ρ is density and B_ϕ is toroidal magnetic field structure), unless there is a permanent equipartition toroidal magnetic field structure. Interestingly, recent helioseismic inferences (Antia 2002; Baldner et al. 2010) from the even degree frequency splittings yield the same order (about $100\,\mathrm{G}$) magnitude of magnetic field structure near the surface.

Another consequence of this near surface decreasing rotation profile is that, during early history of solar system formation, mass from the protoplanetary disk might have accreted on to the Sun and poloidal magnetic field structure might have wound around in that region to form a toroidal magnetic field structure. That means present Sun is slightly more massive and heavy refractory elements (that are supposed to be building blocks of the terrestrial planets in the Sun's vicinity) might have accreted on to the Sun. This in turn means, heavy elemental abundance in the solar core may not be representative of the presently observed surface heavy element abundance as used in the standard solar models and seismic models. This interesting scenario further may solve two well known unsolved solar problems, *viz.*, (1) faint young Sun paradox (for which slightly more massive Sun is required) and, (2) Sun's low angular momentum (as the accretion and transfer of angular momentum to the planetary system are related with each other).

4.5.5 Origin of 22 yr Cycle and Future Helioseismic Inferences

So far understanding the origin of solar cycle and activity phenomena is eluded the scientists and remains one of the important unsolved problem in solar physics. Different variants of Babcok and Leighton and, Parker's *kinematic dynamo* models [supplemented with generation of large scale magnetic field structure from the small scale turbulent fields by Krause (1976) and Radler (1976)] are used to explain successfully some of characteristics of solar cycle and activity phenomena. To recall history of science, helioseismic inferences of rotation rate of the solar convective envelope ruled out the dynamo models that assume radial variation of solar rotation rate profile which increases with the depth. However, contrary to the expectation of kinematic dynamo models, especially from $0.935R_\odot$ downwards, rotation rate decreases.

Recently, flux transport dynamo mechanism (another form of kinematic turbulent dynamo mechanism) that needs meridional circulation flow (single cell flow from surface to pole and return back via base of convection zone and back to the surface), are used to explain the solar cycle periodicity, reversal of magnetic flux, explanation for occurrence of Maunder minimum (occurrence of dearth of sunspot activity, probably on century scales), prediction of future solar cycles, etc. Unfortunately, these flux transport dynamo models, predicted different amplitudes of the present solar cycle. From time to time, consistencies of turbulent kinematic dynamo models are questioned (Piddington 1972; Hiremath 1994 and references there in; Petrovay 2000; Duvall 2001; Venkatakrishnan 2003; Jouve and Brun 2007; Garaud and Brummell 2008; Hiremath 2008; Cally 2009; Hiremath 2009 and references there in; Hiremath 2010 and references there in; Sakurai 2012; Priest 2014; Hathaway 2015; Brun et al. 2015). However, consistent way of understanding the solar cycle and activity phenomena is solution of MHD equations. These solutions (Miesch et al. 2000; Brun and Toomre 2002; Miesch et al. 2008; Miesch and Brown 2012; Brun et al. 2015 and references there in) are compatible with the helioseismically inferred rotation rate (Kosovichev 2011) and characteristics of the solar cycle and, activity phenomena. However, these MHD models, yield many meridional flow cells in the convective envelope contradicting the expectation of the flux transport dynamo models. Hence, at this crucial juncture of solar physics, further helioseismic inferences regarding correct and unambiguous estimation of meridional flow velocity in the convective envelope is necessary. Moreover, magnetic models that are consistent with physics of the solar subsurface and which can explain many observations (11 yr sunspot or 22 yr magnetic cycle, butterfly diagram, polarity reversal, in a particular cycle poleward and equatorward migration of activity phenomena, even and odd degree magnetic oscillations, genesis of polar faculae and coronal holes and their phase difference with the sunspot activity, etc.,) are presently needed.

Acknowledgements I am very much thankful to Dr. Rozelot for the useful suggestions and comments that improved this chapter into a good shape. I am also thankful to Shashanka Gurumath for helping in making corrections to the reference list.

References

Ambroz, P., Druckmller, A. A., Galal, A. A., & Hamid, R.H. (2009). 3D coronal structures and magnetic field during the total solar eclipse of 29 March 2006. *Solar Physics, 258*, 243.

Andersen, B. N. (1996). Theoretical amplitudes of solar g-modes. *Astronomy and Astrophysics, 312*, 610.

Ando, H. (1986). Resonant excitation of the solar g-modes through coupling of 5-min oscillations. *Astrophysics and Space Science, 118*, 177.

Antia, H. M. (2002). Subsurface magnetic fields from helioseismology. In *Proceedings of IAU Coll. 188, ESA SP-505* (p. 71).

Antia, H. M., & Basu, S. (1994). Nonasymptotic helioseismic inversion for solar structure. *Astronomy and Astrophysics, Supplement Series, 107*, 421.

Antia, H. M., Basu, S., & Chitre, S. M. (1998). Solar internal rotation rate and the latitudinal variation of the tachocline. *Monthly Notices of the Royal Astronomical Society, 298*, 543.

Antia, H. M., Basu, S., Pintar, J., & Pohl, B. (2000). Solar cycle variation in solar f-mode frequencies and radius. *Solar Physics, 192*, 459.

Antia, H. M., & Chitre, S. M. (1999). Limits on the proton-proton reaction cross-section from helioseismology. *Astronomy and Astrophysics, 347*, 1000.

Antia, H. M., Chitre, S. M., & Thompson, M. J. (2000). The Sun/s acoustic asphericity and magnetic fields in the solar convection zone. *Astronomy and Astrophysics, 360*, 335.

Appourchaux, T., & Palle, P. (2013). The history of the g-mode quest. In K. Jain, S. C. Tripathy, F. Hill, J. W. Leibacher, & A. A. Pevtsov (Eds.), *ASP series* (Vol. 478, p. 125).

Asplund, M., Grevesse, N., Sauval, A. J., Allende Prieto, C., & Kiselman, D. (2004). Line formation in solar granulation. IV. [O I], O I and OH lines and the photospheric O abundance. *Astronomy and Astrophysics, 417*, 751.

Asplund, M., Grevesse, N., Sauval, A. J., Jacques, S.A., & Scott, P. (2009). The chemical composition of the sun. *Annual Review of Astronomy & Astrophysics, 47*, 481.

Babcock, H. W. (1947). Zeeman effect in stellar spectra. *The Astrophysical Journal, 105*, 105.

Babcock, H. W. (1953). The solar magnetograph. *The Astrophysical Journal, 118*, 387.

Babcock, H. W. (1961). The topology of the sun/s magnetic field and the 22-year cycle. *The Astrophysical Journal, 133*, 572.

Baldner, C. S., Antia, H. M., Basu, S., & Larson, T. P. (2010). Internal magnetic fields inferred from helioseismic data. *Astronomische Nachrichten, 331*, 879.

Basu, S. (1997). Seismology of the base of the solar convection zone. *Monthly Notices of the Royal Astronomical Society, 288*, 572.

Basu, S. (2010). Helioseismology as a diagnostic of the solar interior. *Astrophysics and Space Science, 328*, 43.

Basu, S., & Antia, H. M. (2004). Constraining solar abundances using helioseismology. *The Astrophysical Journal, 606*, 85.

Basu, S., & Antia, H. M. (2008). Helioseismology and solar abundances. *Physics Reports, 457*, 217.

Basu, S., Chaplin, W. J., Elsworth, Y., New, R., & Serenelli A .M. (2009). Fresh insights on the structure of the solar core. *The Astrophysical Journal, 699*, 1403.

Basu, S., Grevesse, N., Mathis, S., & Turck-Chièze S. (2014). Understanding the internal chemical composition and physical processes of the solar interior. *Space Science Reviews*. Online First. doi:10.1007/s11214-014-0035-9.

Belkacem, K., Samadi, R., Goupil, M. J., Dupret, M.A., Brun, A.S., & Baudin, F. (2009). Stochastic excitation of nonradial modes. II. Are solar asymptotic gravity modes detectable? *Astronomy and Astrophysics, 494*, 191.

Böhm-Vitense, E. (1958). Über die Wasserstoffkonvektionszone in Sternen verschiedener Effektivtemperaturen und Leuchtkräfte. Mit 5 Textabbildungen. *Zeitschrift für Astrophysik, 46*, 108.

Brandenburg, A., & Subramaniyam, K. (2005). Astrophysical magnetic fields and nonlinear dynamo theory. *Physics Reports, 417*, 1.

Brun, A. S., Garcia, R. A., Houdek, G., Nandy, D., & Pinsonneault, M. (2015). Erratum to: The solar-stellar connection. *Space Science Reviews*. Online First. doi:10.1007/s11214-015-0157-8.

Brun, A. S., Miesch, M. S., & Toomre, J. (2004). Global-scale turbulent convection and magnetic dynamo action in the solar envelope. *The Astrophysical Journal, 614*, 1073.

Brun, A. S., Miesch, M. S., & Toomre, J. (2011). Modeling the dynamical coupling of solar convection with the radiative interior. *The Astrophysical Journal, 742*, 79.

Brun, A. S., & Toomre, J. (2002). Turbulent convection under the influence of rotation: Sustaining a strong differential rotation. *The Astrophysical Journal, 570*, 865.

Burgess, C. P., Dzhalilov, N. S., Rashba, T. I., Semikoz, V.B., & Valle, J. W. F. (2004). Resonant origin for density fluctuations deep within the sun: Helioseismology and magneto-gravity waves. *Monthly Notices of the Royal Astronomical Society, 348,* 609.

Cally, P. S. (2009). Seismology and the dynamo: History and prospects. In *ASP Conference Series* (Vol. 416, p. 3).

Casanellas, J., & Lopes, I. (2014). The sun and stars: Giving light to dark matter. *Modern Physics Letters A, 29,* 1440001.

Chandrasekhar, S. (1964). A general variational principle governing the radial and the non-radial oscillations of gaseous masses. *The Astrophysical Journal, 139,* 664.

Chaplin, W. J., & Basu, S. (2008). Perspectives in global helioseismology and the road ahead. *Solar Physics, 251,* 53.

Chaplin, W. J., Elsworth, Y., Isaak, G. R., Marchenkov, K.I., Miller, B.A., & New, R. (2001). Changes to low-ll solar p-mode frequencies over the solar cycle: Correlations on different time-scales. *Monthly Notices of the Royal Astronomical Society, 322,* 22.

Charbonneau, P. (2010). Dynamo models of the solar cycle. *Living Reviews in Solar Physics, 7,* 3.

Choudhuri, A. R. (2008). How far are we from a "Standard Model" of the solar dynamo? *Advances in Space Research, 41,* 868.

Choudhuri, A. R., & Gilman, P. A. (1987). The influence of the Coriolis force on flux tubes rising through the solar convection zone. *The Astrophysical Journal, 316,* 788.

Cowling, T. G. (1953). Solar electrodynamics. In G. P. Kaiper (Ed.), *The sun* (p. 532). Chicago: University of Chicago Press.

Dalsgaard, C. (2002). Helioseismology. *Reviews of Modern Physics, 74,* 1073.

Dalsgaard, C. (2003). Lecture Notes on Stellar Oscillations (p. 66). http://w.astro.berkeley.edu/~eliot/Astro202/2009_Dalsgaard.pdf.

Dalsgaard, C., Dappen, W., Ajukov, S. V., Anderson, E. R., Antia, H. M., Basu, S., et al. (1996). The current state of solar modeling. *Science, 272,* 1286.

Damiani, C., Rozelot, J. P., Lefebvre, S., Kilcik, A., & Kosovichev, A. G. (2011). A brief history of the solar oblateness. A review. *Journal of Atmospheric and Solar-Terrestrial Physics, 73,* 241.

D'Silva, S., & Choudhuri, A. R. (1993). A theoretical model for tilts of bipolar magnetic regions. *Astronomy and Astrophysics, 272,* 621.

D'Silva, S., & Howard, R. (1994). Sunspot rotation and the field strengths of subsurface flux tubes. *Solar Physics, 151,* 213.

DeRosa, M. L., Brun, A. S., & Hoeksema, J. T. (2012). Solar magnetic field reversals and the role of dynamo families. *The Astrophysical Journal, 757,* 96.

Deubner, F. -L. (1975). Observations of low wavenumber nonradial eigenmodes of the sun. *Astronomy and Astrophysics, 44,* 371.

Dikpati, M. (2005). Solar magnetic fields and the dynamo theory. *Advances in Space Research, 35,* 322.

Dintrans, B., Brandenburg, A., Nordlund, Å., & Stein, R. F. (2005). Spectrum and amplitudes of internal gravity waves excited by penetrative convection in solar-type stars. *Astronomy and Astrophysics, 438,* 365.

Dubrulle, B., & Knobloch, E. (1993). On instabilities in magnetized accretion disks. *Astronomy and Astrophysics, 274,* 667.

Duvall, T. L. (2001). Observational constraints on solar dynamo models: Helioseismic inferences and magnetic properties. In *AGU Fall Meeting, Abstract # GP22B-03.*

Dziembowski, W. (1983). Resonant coupling between solar gravity modes. *Solar Physics, 82,* 259.

Dziembowski, W. A., & Goode, P. R. (1989). The toroidal magnetic field inside the sun. *The Astrophysical Journal, 347,* 540.

Dziembowski, W. A., Goode, P. R., & Libbrecht, K. G. (1989). The radial gradient in the sun/s rotation. *The Astrophysical Journal, 343,* L53.

Dziembowski, W. A., & Goode, P. R. (1991). Seismology for the fine structure in the sun/s oscillations varying with its activity cycle. *The Astrophysical Journal, 376,* 782.

Dziembowski, W. A., Paterno, L., & Ventura, R. (1985). Excitation of solar oscillation gravity modes by magnetic torque. *Astronomy and Astrophysics, 151,* 47.

Eddy, J. A. (1976). The Maunder Minimum. *Science, 192*, 1189.

Elliott, J. R., Miesch, M. S., & Toomre, J. (2000). Turbulent solar convection and its coupling with rotation: The effect of prandtl number and thermal boundary conditions on the resulting differential rotation. *The Astrophysical Journal, 533*, 546.

Garaud, P., & Brummell, N. H. (2008). On the penetration of meridional circulation below the solar convection zone. *The Astrophysical Journal, 674*, 498.

Garcia, R., Turck-Chièze, S., Jiménez-Reyes, S J., Ballot, J., Pallé, P.L., Eff-Darwich, A., et al. (2007). Tracking solar gravity modes: The dynamics of the solar core. *Science, 316*, 1591.

Gokhale, M. H., & Javaraiah, J. (1992). Global modes constituting the solar magnetic cycle. II - Phases, 'geometrical eigenmodes', and coupling of field behaviour in different latitudes. *Solar Physics, 138*, 399.

Gokhale, M. H., Javaraiah, J., & Hiremath, K. M. (1990). Study of Sun/s "hydromagnetic" oscillations using sunspot data. In J. O. Stenflo (Ed.), *IAU Symp* (Vol. 138, p. 375).

Gough, D. O. (1985). Theory of solar oscillations, Tech. rep. In *ESA future missions in solar, heliospheric and space plasma physics* (pp. 183–197).

Gough, D. O., & Taylor, P. P. (1984). Influence of rotation and magnetic fields on stellar oscillation eigenfrequencies. *Memorie della Societa Astronomica Italiana, 55*, 215.

Gough, D. O., & Thompson, M. J. (1991). The inversion problem. In *Solar interior and atmosphere* (p. 519). Tucson, AZ: University of Arizona Press.

Grevesse, S., & Noels, A. (1993). In N. Prantzos, E. Vangioni-Flam, & M. Casse (Eds.), *Origin and evolution of the elements* (p. 15). Cambridge: Cambridge University Press.

Guenther, D. B., & Demarque, P. (1984). Resonant three-wave interactions of solar g-modes. *The Astrophysical Journal, 277*, L17.

Hale, G. E. (1908). On the probable existence of a magnetic field in sun-spots. *The Astrophysical Journal, 28*, 315.

Hale, G. E. (1913). Preliminary results of an attempt to detect the general magnetic field of the sun. *The Astrophysical Journal, 38*, 27.

Hale, G. E., Seares, F. H., van Maanen, A., & Ellerman, F. (1918). The general magnetic field of the sun. Apparent variation of field-strength with level in the solar atmosphere. *The Astrophysical Journal, 47*, 206.

Hathaway, D. H. (2015). The solar cycle. *Living Reviews in Solar Physics, 12*, 4.

Hiremath, K. M. (1994). Study of Sun/s long period oscillation. Ph.D thesis, Bangalore University.

Hiremath, K. M. (2001). Steady part of rotation and toroidal component of magnetic field in the solar convective envelope. *Bulletin of the Astronomical Society of India, 29*, 169.

Hiremath, K. M. (2006a). Influence of solar activity on the rainfall over India. In N. Gopalswamy, & A. Bhattacharyya (Eds.), *Proceedings of the ILWS Workshop* (p. 178).

Hiremath, K. M. (2006b). The influence of solar activity on the rainfall over India: Cycle-to-cycle variations. *Journal of Astrophysics and Astronomy, 27*, 367.

Hiremath, K. M. (2008). Prediction of solar cycle 24 and beyond. *Astrophysics and Space Science, 314*, 45.

Hiremath, K. M. (2009). Solar forcing on the changing climate. *Sun and Geosphere, 4*, 16.

Hiremath, K. M. (2010). Physics of the solar cycle: New views. *Sun and Geosphere, 5*, 17.

Hiremath, K. M. (2013). Seismology of the sun: Inference of thermal, dynamic and magnetic field structures of the interior. In M. Mohan (Ed.), *New trends in atomic and molecular physics*. Springer series on atomic, optical, and plasma physics (pp. 332–333). Berlin/Heidelberg: Springer.

Hiremath, K. M., & Gokhale, M. H. (1995). "Steady" and "fluctuating" parts of the sun's internal magnetic field: Improved model. *The Astrophysical Journal, 448*, 437.

Hiremath, K. M., Hegde, M., & Soon, W. (2015). Indian summer monsoon rainfall: Dancing with the tunes of the sun. *New Astronomy, 35*, 8.

Hiremath, K. M., & Lovely, M. R. (2007). Magnetic flux in the solar convective envelope inferred from initial observations of sunspots. *The Astrophysical Journal, 667*, 585.

Hiremath, K. M., & Lovely, M. R. (2012). Toroidal magnetic field structure of the solar convective envelope inferred from the initial observations of the sunspots. *New Astronomy, 17*, 392.

Hiremath, K. M., & Mandi, P. I. (2004). Influence of the solar activity on the Indian Monsoon rainfall. *New Astronomy, 9*, 651.

Howe, R., Komm, R., & Hill, F. (1999). Solar cycle changes in GONG P-mode frequencies, 1995–1998. *The Astrophysical Journal, 524*, 1084.

Ikhsanov, R. N., & Tavastsherna, K. S. (2013). High-latitude coronal holes and polar faculae in the 21st-23rd solar activity cycles. *Geomagnetism and Aeronomy, 53*, 896.

Jimenez, A., Garcia, R. A., & Palle, P. L. (2011). The acoustic cutoff frequency of the sun and the solar magnetic activity cycle. *The Astrophysical Journal, 743*, 99.

Jouve, L., Brun, A. S. (2007). On the role of meridional flows in flux transport dynamo models. *Astronomy and Astrophysics, 474*, 239.

Karna, N., Hess Webber, S. A., & Pesnell, W.D. (2014). Using polar coronal hole area measurements to determine the solar polar magnetic field reversal in solar cycle 24. *Solar Physics, 289*, 3381.

Kitchatinov, L. L., & Rüdiger, G. (1995). Differential rotation in solar-type stars: Revisiting the Taylor-number puzzle. *Astronomy and Astrophysics, 299*, 446.

Knaack, R., & Stenflo, J. O. (2005). Spherical harmonic decomposition of solar magnetic fields. *Astronomy and Astrophysics, 438*, 349.

Kosovichev, A. G. (2011). Advances in global and local helioseismology: An introductory review. In *The pulsations of the sun and the stars*. Lecture notes in physics (Vol. 832, pp. 3–84). Heidelberg: Springer.

Kosovichev, A. G., Christensen-Dalsgaard, J., Daeppen, W., Dziembowski, W.A., Gough, D.O., & Thompson, M.J. (1992). Sources of uncertainty in direct seismological measurements of the solar helium abundance. *Monthly Notices of the Royal Astronomical Society, 259*, 536.

Krause, F. (1976). Mean-field magnetohydrodynamics of the solar convection zone, in mechanisms of solar activity. In V. Bumba, & J. Kleczek (Eds.), *Proceedings from IAU Symposium* (Vol. 71, p. 305).

Kuhn, J. R. (1988). Helioseismological splitting measurements and the nonspherical solar temperature structure. *The Astrophysical Journal, 331*, L131.

Kuker, M., Rudiger, G., & Kitchatinov, L. L. (2011). The differential rotation of G dwarfs. *Astronomy and Astrophysics, 530*, 48.

Kumar, P., Quataert, E. J., & Bahcall, J. N. (1996). Observational searches for solar g-modes: Some theoretical considerations. *The Astrophysical Journal, 458*, 83.

Lefebvre, S., & Kosovichev, A. G. (2005). Changes in the subsurface stratification of the sun with the 11-year activity cycle. *The Astrophysical Journal, 633*, 149.

Lefebvre, S., Kosovichev, A. G., & Rozelot, J. (2007). Helioseismic test of nonhomologous solar radius changes with the 11 year activity cycle. *The Astrophysical Journal, 658*, 135L.

Lefebvre, S., Nghiem, P. A., & Turck-Chièze, S. (2007). Impact of a radius and composition variation on stratification of the solar subsurface layers. *The Astrophysical Journal, 690*, 1272L.

Leibacher, J. W. (1971). Solar atmospheric oscillations. Ph.D thesis, Harvard University.

Leighton, R. B. (1964). Transport of magnetic fields on the sun. *The Astrophysical Journal, 140*, 1547.

Leighton, R. B. (1969). A magneto-kinematic model of the solar cycle. *The Astrophysical Journal, 156*, 1.

Lopes, I., Panci, P., & Silk, J. (2014). Helioseismology with Long-range Dark Matter-Baryon Interactions. *The Astrophysical Journal, 795*, 11.

Lopes, I., & Silk, J. (2014). Helioseismology and asteroseismology: Looking for gravitational waves in acoustic oscillations. *The Astrophysical Journal, 794*, 32.

Lynden-Bell, D., & Ostriker, J. P. (1967). On the stability of differentially rotating bodies. *Monthly Notices of the Royal Astronomical Society, 136*, 293.

Maunder, E. W. (1904). Note on the distribution of sun-spots in heliographic latitude, 1874–1902. *Monthly Notices of the Royal Astronomical Society, 64*, 747.

McKernan, B., Ford, K. E. S., Kocsis, B., & Haiman, Z. (2014). Stars as resonant absorbers of gravitational waves. *Monthly Notices of the Royal Astronomical Society, 445L*, 74.

Meftah, M., Irbah, A., Hauchecorne, A., Corbard, T., Turck-Chièze, S., Hochedez, J.F., et al. (2015). On the determination and constancy of the solar oblateness. *Solar Physics, 290*, 673.

Mestel, L. (Ed.) (1999). *Stellar magnetism* (p. 395). Oxford: Clarendon.

Miesch, M. S. (2012). The solar dynamo. *Philosophical Transactions of the Royal Society A: Mathematical, Physical and Engineering Sciences, 370*, 3049.

Miesch, M. S., & Brown, B. P. (2012). Convective Babcock-Leighton dynamo models. *The Astrophysical Journal, 746*, 26.

Miesch, M. S., Brun, A. S., De Rosa, M. L., & Toomre, J. et al. (2008). Structure and evolution of giant cells in global models of solar convection. *The Astrophysical Journal, 673*, 557.

Miesch, M. S., Brun, A. S., & Toomre, J. (2006). Solar differential rotation influenced by latitudinal entropy variations in the tachocline. *The Astrophysical Journal, 641*, 618.

Miesch, M. S., Elliott, J. R., Toomre, J., Clune, T.L., Glatzmaier, G.A., & Gilman, P.A. (2000). Three-dimensional spherical simulations of solar convection. I. Differential rotation and pattern evolution achieved with laminar and turbulent states. *The Astrophysical Journal, 532*, 593.

Mordinov, A. V., & Yazov, S. A. (2014). Reversals of the Sun's Polar Magnetic Fields in Relation to Activity Complexes and Coronal Holes. *Solar Physics, 289*, 1971.

Ossendrijver, M. (2003). The solar dynamo. *Astronomy and Astrophysics Review, 11*, 287.

Pasachoff, J. M. (2009). Solar eclipses as an astrophysical laboratory. *Nature, 459*, 789.

Pasachoff, J. M., Ruin, V., Druckmller, M., Aniol, P., Saniga, M., & Minarovjech, M. (2009). The 2008 August 1 eclipse solar-minimum corona unraveled. *The Astrophysical Journal, 702*, 1297.

Paterno, L., Sofia, S., & di Mauro, M. P. (1996). The rotation of the Sun's core. *Astronomy and Astrophysics, 314*, 940.

Petrovay, K. (2000). What makes the sun tick? The origin of the solar cycle. In *ESA SP* (Vol. 463, p. 3).

Piddington, J. H. (1972). Solar dynamo theory and the models of babcock and leighton. *Solar Physics, 22*, 3.

Priest, E. (Ed.) (2014). In *Magnetohydrodynamics of the sun* (pp. 304–305). Cambridge: Cambridge University Press.

Rabello-Soares, M. C. (2012). Solar-cycle variation of sound speed near the solar surface. *The Astrophysical Journal, 745*, 184.

Rabello-Soares, M. C., Korzennik, S. G., & Schou, J. (2008). Analysis of MDI high-degree mode frequencies and their rotational splittings. *Solar Physics, 251*, 197.

Radler, K. H. (1976). Mean-field magnetohydrodynamics as a basis of solar dynamo theory. In V. Bumba, & J. Kleczek (Eds.), *Proceedings from IAU Symposium* (Vol. 71, p. 305).

Rashba, T. (2008). Probing the internal solar magnetic field through g-modes. *Journal of Physics: Conference Series, 118*, 012085.

Rashba, T. I., Semikoz, V. B., Turck-Chièze, S., & Valle, J. W. F. (2007). Probing the internal solar magnetic field through g modes. *Monthly Notices of the Royal Astronomical Society, 377*, 453.

Rashba, T. I., Semikoz, V. B., & Valle, J. W. F. (2006). Radiative zone solar magnetic fields and g modes. *Monthly Notices of the Royal Astronomical Society, 370*, 845.

Rempel, M. (2005). Solar differential rotation and meridional flow: The role of a subadiabatic tachocline for the taylor-proudman balance. *The Astrophysical Journal, 622*, 1320.

Robinson, F. J., & Chan, K. L. (2001). A large-eddy simulation of turbulent compressible convection: Differential rotation in the solar convection zone. *Monthly Notices of the Royal Astronomical Society, 321*, 723.

Ronan, R. S., Cadora, K., & Labonte, B. J. (1994). Solar cycle changes in the high frequency spectrum. *Solar Physics, 150*, 389.

Rozelot, J. P., & Damiani, C. (2011). History of solar oblateness measurements and interpretation. *The European Physical Journal H, 36*, 407.

Rozelot, J.-P., Damiani, C., Kilcik, A., Tayoglu, B., & Lefebvre, S. (2011). Unveiling stellar cores and multipole moments via their flattening. In *The pulsations of the sun and the stars*. Lecture notes in physics (Vol. 832, p. 161). Heidelberg: Springer.

Rozelot, J. P., & Fazel, Z. (2013). Revisiting the solar oblateness: Is relevant astrophysics possible? *Solar Physics, 287*, 161.

Rozelot, J. P., & Lefebvre, S. (2006). In J.-P. Rozelot (Ed.), *Solar and heliospheric origins of space weather phenomena* (p. 5). New York, NY: LLC.

Sakurai, T. (2012). Helioseismology, dynamo, and magnetic helicity. In H. Shibahashi, M. Takata, & A.E. Lynas-Gray (Esd.), *ASP Conference Proceeding* (Vol. 462, p. 247). San Francisco: Astronomical Society of the Pacific.

Severnyi, A. B., Kotov, V. A., & Tsap, T. T. (1976). Observations of solar pulsations. *Nature, 259*, 87.

Shibahashi, H., Hiremath, K. M., & Takata, M. (1998a). Seismic model of the solar convective envelope. In *ESASP* (Vol. 418, p. 537).

Shibahashi, H., Hiremath, K. M., & Takata, M. (1998b). A seismic model of the solar convective envelope. In *IAU Symp* (Vol. 185, p. 81).

Shibahashi, H., Hiremath, K. M., & Takata, M. (1999). Seismic solar model: Both of the radiative core and the convective envelope. *Advances in Space Research, 24*, 177.

Shibahashi, H., & Takata, M. (1996). A seismic solar model deduced from the sound-speed distribution and an estimate of the neutrino fluxes. *Publications of the Astronomical Society of Japan, 48*, 377.

Shibahashi, H., & Takata, M. (1997). A seismic solar model deduced from the sound speed distribution and an estimate of the neutrino fluxes. In *PASJ, IAU Symp, 181*, 167.

Shibahashi, H., Takata, M., & Tanuma, S. (1995). A seismic solar model deduced from the sound speed distribution. In *ESA SP, Proceedings of the 4th Soho Workshop* (p. 9).

Siegel, D. M., & Markus, R. (2014). An upper bound from helioseismology on the stochastic background of gravitational waves. *The Astrophysical Journal, 784*, 88.

Solanki, S., Inhester, B., & Schussler, M. (2006). The solar magnetic field. *Reports on Progress in Physics, 69*, 563.

Stenflo, J. O. (1988). Global wave patterns in the sun's magnetic field. *Astrophysics and Space Science, 144*, 321.

Stenflo, J. O. (1993). Cycle patterns of the axisymmetric magnetic field. In R. J. Rutten, & C.J. Schrijver (Eds.), *Solar surface magnetism*. NATO advanced science institutes (ASI) series (p. 365).

Stenflo, J. O., & Vogel, M. (1986). Global resonances in the evolution of solar magnetic fields. *Nature, 319*, 285.

Stix, M. (1972). Non-linear dynamo waves. *Astronomy and Astrophysics, 20*, 9.

Sturrock, P. A., & Gilvarry, J. J. (1967). Solar oblateness and magnetic field. *Nature, 216*, 1280.

Takata, M., & Shibahashi, H. (1998). Solar models based on helioseismology and the solar neutrino problem. *The Astrophysical Journal, 504*, 1035.

Tripathy, S. C., Jain, K., & Hill, F. (2013). Acoustic mode frequencies of the sun during the minimum phase between solar cycles 23 and 24. *Solar Physics, 282*, 1.

Turck-Chièze, S., Garcia, R., Lopes, I., Ballot, J., Couvidat, S., Mathur, S., et al. (2012). First study of dark matter properties with detected solar gravity modes and neutrinos. *The Astrophysical Journal, 746L*, 12.

Turck-Chièze, S., & Lopes, I. (2012). Solar-stellar astrophysics and dark matter. *Research in Astronomy and Astrophysics, 12*, 1107.

Ulrich, R. K. (1970). The five-minute oscillations on the solar surface. *The Astrophysical Journal, 162*, 993.

Unno, W., Osaki, Y., Ando, H., Sao, H., & Shibahashi, H. (1989). *Noradial oscillations of stars* (p. 108, 2nd ed.). Tokyo: University of Tokyo Press.

Venkatakrishnan, P. (2003). Solar magnetic fields. In H. M. Antia, A. Bhatnagar, & P. Ulmschnei-der (Eds.), *Lecture notes in physics* (Vol. 619, p. 202). Heidelberg: Springer.

Vincent, A. C., Scott, P., & Serenelli, A. (2015). Possible indication of momentum-dependent asymmetric dark matter in the sun. *Physical Review Letters, 114*, 081302.

Wentzel, D. G. (1987). Solar oscillations - Generation of a g-mode by two p-modes. *The Astrophysical Journal, 319*, 966.

Wolff, C. L., & O'Donovan, A. E. (2007). Coupled groups of g-modes in a sun with a mixed core. *The Astrophysical Journal, 661*, 568.

Woodard, M. (1984). Upper limit on solar interior rotation. *Nature, 309*, 530.

Woodard, M. F., & Noyes, R. W. (1985). Change of solar oscillation eigenfrequencies with the solar cycle. *Nature, 318*, 449.

Xu, Y., Yang, R., Zhang, Q., & Xu, G. (2011). Solar oblateness and Mercury's perihelion precession. *Monthly Notices of the Royal Astronomical Society, 415*, 3335.

Yoshimura, H. (1978). Nonlinear astrophysical dynamos - Multiple-period dynamo wave oscillations and long-term modulations of the 22 year solar cycle. *The Astrophysical Journal, 220*, 706.

Chapter 5
Physical Processes Leading to Surface Inhomogeneities: The Case of Rotation

Michel Rieutord

Abstract In this lecture I discuss the bulk surface heterogeneity of rotating stars, namely gravity darkening. I especially detail the derivation of the ω-model of Espinosa Lara and Rieutord (Astron Astrophys 533:A43, 2011), which gives the gravity darkening in early-type stars. I also discuss the problem of deriving gravity darkening in stars owning a convective envelope and in those that are members of a binary system.

5.1 Introduction

As it has been much discussed in this school, surface inhomogeneities of stars are more and more frequently detected due to the increasing sensitivity of the instruments. If correctly understood, and therefore modeled, these data may open new windows on the interior or on the history of stars.

The purpose of this lecture is to first briefly review the processes that lead to such inhomogeneities and then to focus on a very fundamental one, namely rotation.

In the ancient times, when the eye was the only optical instrument to observe Nature, the Sun was thought as a pure uniform bright disc. The invention of the telescope by Galileo ruined this idea, showing that the Sun was spherical with spots on it. Presently, it suffices to have a look at images of the Sun taken at short wavelengths to understand that its surface is certainly not uniform. Such images actually reveal that the magnetic fields are a prominent cause of this non-uniformity. It looks like a mess which even impacts the distribution of surface flux and temperature. A closer look at the magnetic structures but also below, at the photospheric level, shows that all these heterogeneities evolve with time. On the photosphere, turbulent convection features the surface with two important scales: granulation and supergranulation (Rieutord and Rincon 2010). Even the bulk surface rotation is not uniform. This differential rotation, known since the nineteenth

M. Rieutord (✉)
Université de Toulouse, UPS-OMP, IRAP, Toulouse, France

CNRS, IRAP, 14, Avenue Edouard Belin, F-31400 Toulouse, France
e-mail: mrieutord@irap.omp.eu

© Springer International Publishing Switzerland 2016
J.-P. Rozelot, C. Neiner (eds.), *Cartography of the Sun and the Stars*,
Lecture Notes in Physics 914, DOI 10.1007/978-3-319-24151-7_5

century, with fast equatorial regions and slow polar regions, is now understood as driven by Reynolds stresses coming from the turbulence in the solar convection zone.

Hence, the surface of the Sun teaches us that we should expect non-uniform velocities, temperatures, flux and magnetic fields at the surface of all low-mass stars. But one consequence of the strong mixing imposed by turbulent convection and the ever changing magnetic fields is that the solar photosphere has a uniform chemical composition!

But uniform chemical composition is certainly not possible when turbulent convection disappears and no longer mixes the surface layers, that is when we consider stars of higher mass with an outer radiative envelope. There, combination of magnetic fields with microscopic diffusion processes (gravitational settling or radiative acceleration) may on the contrary lead to chemical surface inhomogeneities (Vauclair and Vauclair 1982; Alecian 2013; Korhonen et al. 2013). But even when magnetic fields are absent at their surface, early-type stars are still endowed with a non-uniform surface: absence of magnetic field is indeed correlated with fast rotation, a feature that makes the polar caps brighter than the equatorial regions.

From the foregoing presentation we see that three processes, intrinsic to each star may lead to surface inhomogeneities: rotation, convection and magnetic fields. There is a fourth one, but extrinsic to the star itself, namely binarity. A companion indeed raises tides, illuminates one side of the star or may even transfer mass.

Within these four physical processes that make the surface of stars not uniform, we shall concentrate on the most simple, namely rotation, which, a priori leads surface variations that only depend on the latitude. We shall discuss in detail this very basic physical effect, leaning on the recent works of Espinosa Lara and Rieutord (2011, 2012).

5.2 The Energy Flux in Radiative Envelopes of Rotating Stars

5.2.1 *von Zeipel 1924*

In a seminal paper, von Zeipel (1924) showed that a rotating star may be brighter at the poles than at the equator. This result is quite simple to derive if we assume that the star is barotropic and that the energy flux is given by Fourier's law. Indeed, if the star is barotropic, meaning that its equation of state can be simplified to

$$P \equiv P(\rho),$$

it implies that there exist a hydrostatic solution in the rotating frame and that all thermodynamic quantities can be expressed as a function of the total potential Φ

(i.e. gravitational plus centrifugal). Hence, one writes

$$\rho \equiv \rho(\Phi), \quad T \equiv T(\Phi), \quad \text{etc.}$$

Then, using Fourier's law to derive the heat flux, one finds

$$\boldsymbol{F}_{\text{rad}} = -\chi \nabla T = -\chi(\Phi) T'(\Phi) \nabla \Phi = K(\Phi) \boldsymbol{g}_{\text{eff}}$$

whence von Zeipel law

$$T_{\text{eff}} = K g_{\text{eff}}^{1/4} \quad \text{on the surface } \Phi = \text{Cst}$$

This result is simple but incorrect. Indeed, barotropic stars are realized in two cases: either the star is isentropic and thus fully convective or it is isothermal but this can hardly be the case.[1] So the closest case that may match the barotropic state is that of a fully convective star, but in such a case the flux cannot be derived from the Fourier's law. In fact, these hypothesis (barotropicity and heat diffusion) lead to a contradiction. Indeed, we may note that the total potential and the temperature would verify in the envelope of the star,

$$\begin{cases} \text{Div}(\chi \nabla T) = 0 \\ \Delta \Phi = 4\pi G \rho + 2\Omega^2 \end{cases} \tag{5.1}$$

which leads to

$$\text{Div}(\chi(\Phi) T'(\Phi) \nabla \Phi) = 0 \quad \Longleftrightarrow \quad 4\pi G \rho + 2\Omega^2 + (\ln(\chi T'))' g_{\text{eff}}^2 = 0$$

On an equipotential, ρ is constant as well as $(\ln(\chi T'))'$, but g_{eff} is not constant. Hence, the latter equation is impossible. The reason for that is that for a rotating star where heat is transported by diffusion, a barotropic state cannot be and should be replaced by a baroclinic state. In such a state, isobars, isotherms or equipotential are all different, not very different, but different. This is the normal state that comes from the fact (basically) that temperature, pressure, and gravitational potential all obeys different and independent equations. The barotropic state is therefore rather peculiar (but see Rieutord 2006, for a more detailed presentation).

Now one may wonder if it is possible to derive the dependence of the flux with latitude for a rotating star without computing the whole stellar structure and the associated flows as in Espinosa Lara and Rieutord (2013). Fortunately, this is indeed possible as Espinosa Lara and Rieutord (2011) have shown. It is not as simple as von Zeipel law, but it has the merit of relying on controllable hypothesis.

---

[1] Some models of prestellar core use this hypothesis, sometimes.

5.2.2   The Idea of the ω-Model

In the following we shall first restrict ourselves to the case of early-type stars, that
is to stars that have a radiative envelope around a convective core. We'll discuss the
case of convective envelope in the next section.

Within the envelope of a star the flux just obeys:

$$\mathrm{Div} \boldsymbol{F} = 0 \tag{5.2}$$

namely energy is conserved and there are no energy sources.

This is a single equation, not enough to determine the two components, (F_r, F_θ),
of the flux, but if we add a constraint to the flux we may find it. We thus assume that
the flux is anti-parallel to the effective gravity

$$\boldsymbol{F} = -f(r, \theta) \boldsymbol{g}_{\mathrm{eff}} \tag{5.3}$$

In order to avoid an additional unknown, we shall take the effective gravity $\boldsymbol{g}_{\mathrm{eff}}$ as
given by the Roche model. In such a case we shall see that the flux function f can be
determined and that the latitude variation of the flux depends on a single parameter
ω defined as the ratio of the angular velocity to the keplerian angular velocity at the
equator. In other words, the flux depends on

$$\omega = \frac{\Omega}{\Omega_k} = \Omega \left(\sqrt{\frac{GM}{R_e^3}} \right)^{-1} \tag{5.4}$$

Thus, we shall call this model the ω-model to emphasize the crucial role of the
reduced angular velocity ω.

However, before going any further, we may wonder whether the assumptions are
strong or not, especially (5.3).

In a radiative zone, the configuration is baroclinic so vectors are surely not
aligned but fortunately we can now revert to 2D-models to get an idea of the
misalignment. As shown in Fig. 5.1, the misalignment remains small, less than a
degree, even if the star rotates close to criticality.

Thus, even if the envelope is the seat of baroclinic flows, the misalignment
is small. Actually, the baroclinic torque $(\nabla P \times \nabla \rho)/\rho^2$ does not need a strong
misalignment of the vectors to be efficient at driving baroclinic flows because the
two gradients (of pressure and density) are already quite strong.

Let us pursue somewhat. From (5.2) and (5.3), we have

$$\mathrm{Div} \boldsymbol{F} = 0 \quad \Longleftrightarrow \quad \mathrm{Div}(f \nabla \Phi) = 0$$

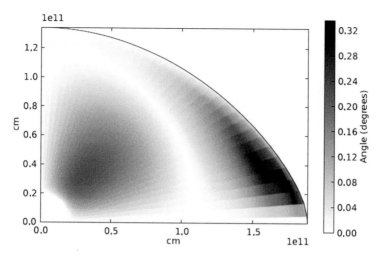

Fig. 5.1 Misalignment between pressure gradient and flux for a configuration with a flatness
∼30 % (from Espinosa Lara and Rieutord 2011)

thus

$$\mathbf{g}_{\text{eff}} \cdot \nabla \ln f = -2\Omega^2 \tag{5.5}$$

because $\Delta\Phi = -2\Omega^2$; hence,

$$\frac{\partial \ln f}{\partial \xi} = -\frac{2\Omega^2}{g_{\text{eff}}} \tag{5.6}$$

where we introduced the local vertical coordinate ξ. Equality (5.6) shows that
$\frac{\partial \ln f}{\partial \xi}$, and therefore f, has latitudinal variations similar to those of g_{eff}. Hence the
horizontal variations of the flux cannot be given by von Zeipel law. In other words
$T_{\text{eff}}/g_{\text{eff}}^{1/4}$ cannot be constant.

The second hypothesis is the use of the Roche model. This model assumes that
the whole mass of the star is concentrated at the centre thus leading to a gravitational
potential in $1/r^2$ everywhere. For the regions we are interested in, namely the
envelope of early-type stars, this is a rather good approximation since these stars
are usually said to be "centrally condensed". In Fig. 5.2 we show the density profile
of a non-rotating $5\,M_\odot$ star along with the profile of two polytropes. We see that
the n = 3 polytrope represents fairly well the density profile of the star and that the
n = 3/2-polytrope, which is a very good model for fully convective stars, is much
less "centrally condensed". Hence, gravity in the outer envelope of an early-type star
is well represented by the Roche model (see Fig. 5.2 right). The interior discrepancy
with the Roche model has no consequence for our purpose.

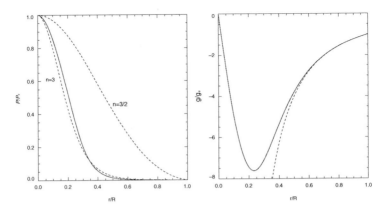

Fig. 5.2 *Left*: Density profile of a M $= 5\,M_\odot$ ZAMS, non rotating star (*solid line*) together with that of a n $= 3/2$ and n $= 3$ polytropes. *Right*: The interior gravity of the same stellar model (*solid line*) together with the $-1/r^2$ Roche model (*dashed line*). The ZAMS model is an ESTER model (e.g. Espinosa Lara and Rieutord 2013)

5.2.3   The Derivation of $f(r, \theta)$

f is given by (5.5) but we first need to scale this function so as to introduce a non-dimensional function F that accounts for the radial and latitudinal variation of the flux. This is easily done if we observe that near the star's centre

$$F \sim \frac{L}{4\pi r^2}e_r \qquad g_{\text{eff}} \sim -\frac{GM}{r^2}e_r$$

So that we may set

$$f(r, \theta) = \frac{L}{4\pi GM}F(r, \theta) \tag{5.7}$$

with

$$\lim_{r \to 0} F(r, \theta) = 1$$

Then, we scale the gravity with GM/R_e^2 and the length scale with the equatorial radius R_e. The scaled angular velocity is therefore given by

$$\omega = \frac{\Omega}{\Omega_k} = \Omega\left(\sqrt{\frac{GM}{R_e^3}}\right)^{-1}$$

At this point we should underline that the angular velocity is scaled by the keplerian angular velocity given by the equatorial radius. It is often the case in the literature that the scale of angular velocity is the critical velocity associated with the Roche model of the considered mass M (e.g. Monnier et al. 2012, for instance). This gives a different ω (i.e. fraction of critical velocity). We give in the Appendix the relation between these two ways of appreciating angular velocity.

We now proceed to the derivation of F the scaled version of f. From (5.7) and (5.5) we get

$$\left(\frac{1}{\omega^2 r^2} - r\sin^2\theta\right)\frac{\partial F}{\partial r} - \sin\theta\cos\theta\frac{\partial F}{\partial\theta} = 2F$$

With $F(0,\theta) = 1$ we have all the elements for solving the equation for F.

First, we solve for $\ln F$, namely,

$$\left(\frac{1}{\omega^2 r^2} - r\sin^2\theta\right)\frac{\partial \ln F}{\partial r} - \sin\theta\cos\theta\frac{\partial \ln F}{\partial\theta} = 2 . \tag{5.8}$$

If we set $\ln F = \ln G + A(\theta)$, so that $A(\theta)$ removes the RHS of (5.8), we immediately find that

$$A'(\theta) = -2/\sin\theta\cos\theta \implies A(\theta) = -\ln(\tan^2\theta) .$$

But we still have to solve the homogeneous equation, namely

$$\left(\frac{1}{\omega^2 r^2} - r\sin^2\theta\right)\frac{\partial \ln G}{\partial r} - \sin\theta\cos\theta\frac{\partial \ln G}{\partial\theta} = 0 \tag{5.9}$$

Such a first order partial differential equations is solved by the method of characteristics. We therefore look for places where $\ln G$ is constant. These places are called the characteristics curves of G. They are such that

$$\frac{\partial \ln G}{\partial r}dr + \frac{\partial \ln G}{\partial\theta}d\theta = 0$$

but G also verifies (5.9) so that we can eliminate $\frac{\partial \ln G}{\partial r}$ and $\frac{\partial \ln G}{\partial\theta}$ and get

$$\left(\frac{1}{\omega^2 r^2} - r\sin^2\theta\right)d\theta + \sin\theta\cos\theta dr = 0 \tag{5.10}$$

which is the equation of characteristics.

We first observe that we may multiply this equation by any function $H(r, \theta)$ without changing anything. So we may also look for h such that

$$\begin{cases} \dfrac{\partial h}{\partial r} = H \sin \theta \cos \theta \\[4mm] \dfrac{\partial h}{\partial \theta} = H \left(\dfrac{1}{\omega^2 r^2} - r \sin^2 \theta \right) \end{cases} \tag{5.11}$$

where H needs to be chosen so that this system can be integrated. After trial and error, we find that $H = \omega^2 r^2 \cos \theta \cot \theta$ is the right function. Thus

$$\begin{cases} \dfrac{\partial h}{\partial r} = \omega^2 r^2 \cos^3 \theta \\[4mm] \dfrac{\partial h}{\partial \theta} = \dfrac{\cos^2 \theta}{\sin \theta} - \omega^2 r^3 \cos^2 \theta \sin \theta \end{cases} \tag{5.12}$$

and the solution is

$$h(r, \theta) = \frac{1}{3} \omega^2 r^3 \cos^3 \theta + \cos \theta + \ln \tan(\theta/2)$$

The curves $h(r, \theta) = $ Cst are the characteristics. Note that the polar equation of a characteristic, namely the dependence $r \equiv r(\theta)$, is just implicitly known, and depends on the chosen constant.

Now, we know that $\ln G$ or G is constant on the curves where $h(r, \theta) = $ Cst. So we can write

$$G \equiv G(h) . \tag{5.13}$$

It means that the variations of G with (r, θ) are through those of $h(r, \theta)$ only. So we find that

$$\ln F = \ln G(h) - \ln \tan^2 \theta \qquad \text{or} \qquad F = \frac{G(h(r, \theta))}{\tan^2 \theta}$$

This is the solution of the partial differential equation, but it is up to an arbitrary function $G(h)$ that we should determine. For that, we need to revert to the boundary conditions, namely that $F(0, \theta) = 1$. We thus need to impose

$$\frac{G(h(0, \theta))}{\tan^2 \theta} = 1 \tag{5.14}$$

or

$$G(\cos \theta + \ln \tan(\theta/2)) = \tan^2 \theta \tag{5.15}$$

for all θ. This is certainly a weird expression of G, but actually it is sufficient. Let's introduce the function h_0 such that

$$h_0(\theta) = \cos\theta + \ln\tan(\theta/2) \tag{5.16}$$

Hence, we have

$$(G \circ h_0)(\theta) = \tan^2\theta$$

or

$$G \circ h_0 = \tan^2 \qquad \Longrightarrow \qquad G = \tan^2 \circ h_0^{-1}$$

so formally, the solution for G is

$$G(r, \theta) = \tan^2(h_0^{-1}(h(r, \theta)))$$

To make it more understandable, we set

$$\psi = h_0^{-1}(h(r, \theta)) \tag{5.17}$$

so that

$$h_0(\psi) = \frac{1}{3}\omega^2 r^3 \cos^3\theta + \cos\theta + \ln\tan(\theta/2)$$

or

$$\cos\psi + \ln\tan(\psi/2) = \frac{1}{3}\omega^2 r^3 \cos^3\theta + \cos\theta + \ln\tan(\theta/2) \tag{5.18}$$

which is a transcendental equation for ψ. However, it is not difficult to solve numerically (we know that when r or ω are small $\psi \simeq \theta$). So finally we find

$$F(r, \theta) = \frac{\tan^2(\psi(r, \theta))}{\tan^2\theta} \tag{5.19}$$

where $\psi(r, \theta)$ is given by (5.18).

5.2.4   Two Interesting Latitudes

F seems to be singular at the pole ($\theta = 0$) and at the equator ($\theta = \pi/2$). Let us explore these two latitudes.

Starting with the pole, we see that if $\theta \ll 1$, then, from (5.18), we find that $\psi \ll 1$ as well. Indeed, for small values of the angles we have

$$1 + \ln \tan(\psi/2) \simeq \frac{1}{3}\omega^2 r^3 + 1 + \ln \tan(\theta/2)$$

so that

$$\psi \simeq \theta e^{\omega^2 r^3/3} \tag{5.20}$$

and

$$F(r,0) = e^{2\omega^2 r^3/3} \tag{5.21}$$

which gives the values of F along the rotation axis.

The equator is more complicated. We need to know that if $\varepsilon \ll 1$ then

$$\ln\left(\tan\left[\frac{\pi}{4} - \varepsilon\right]\right) = -\varepsilon - \frac{1}{6}\varepsilon^3 - \cdots$$

With this asymptotic expansion we find

$$F(r, \pi/2) = (1 - \omega^2 r^3)^{-2/3}$$

5.2.5 The Final Solution of the ω-Model

Back to the definitions we started with, we can express the flux with the effective gravity in the following way:

$$F = -\frac{L}{4\pi GM} F(\omega, r, \theta) g_{\text{eff}} \tag{5.22}$$

so that we also get the effective temperature

$$T_{\text{eff}} = \left(\frac{L}{4\pi\sigma GM}\right)^{1/4} \sqrt{\frac{\tan\psi}{\tan\theta}}\, g_{\text{eff}}^{1/4} \tag{5.23}$$

From this expression, we see that the function $\sqrt{\tan\psi/\tan\theta}$ shows the deviation from the von Zeipel law.

Noting that

$$g_{\text{eff}} = \frac{GM}{R_e^2}\left(-\frac{e_r}{r^2} + \omega^2 r \sin\theta e_s\right)$$

for the Roche model (e_s is the unit radial vector of cylindrical coordinates and e_r that of spherical coordinates). We find that the flux is given by

$$F = -\frac{L}{4\pi R_e^2}\left(-\frac{e_r}{r^2} + \omega^2 r \sin\theta e_s\right) F(\omega, r, \theta) \qquad (5.24)$$

which shows that it depends only on ω and a scaling factor $\frac{L}{4\pi R_e^2}$, hence the name "ω-model".

5.2.6 Comparison with 2D Models: A Test of the β- and ω-Models

After the foregoing mathematical developments we certainly would like to compare the results of this modeling to more elaborated models. For this purpose, we compared the latitude variations of the flux with the prediction of fully two-dimensional ESTER models (Espinosa Lara and Rieutord 2013). We recall that ESTER models give a full solution of the internal structure of a rotating early-type star including the differential rotation and the meridional circulation driven by the baroclinicity of the envelope. They also include the full microphysics (opacity and equation of state) from OPAL tables. Figures 5.3 and 5.4 show that the ω-model matches very well the output of the full ESTER models. Moreover, we also note

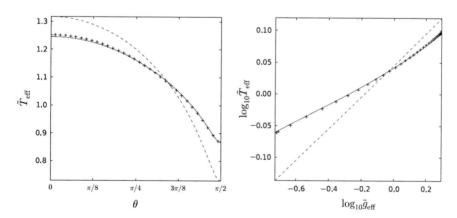

Fig. 5.3 *Left*: Scaled effective temperature as a function of colatitude for a $3\,M_\odot$ model at $\Omega = 0.9\Omega_k$. The *solid line* shows the prediction of the simplified model, 'pluses' show the prediction of a fully 2D ESTER model including differential rotation (see Espinosa Lara and Rieutord 2013, for details), while the *dashed line* is for the von Zeipel law. *Right*: With the same *symbols* as on *left*, the effective temperature as a function of the effective gravity (plots from Espinosa Lara and Rieutord 2011)

Fig. 5.4 Variation of the
ratio of effective temperature
at pole and equator as a
function of the flatness of the
star. *Symbols* are the same as
in Fig. 5.3 (plots from
Espinosa Lara and Rieutord
2011)

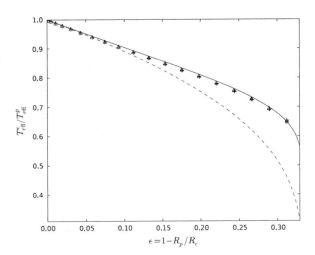

that the dependence of the effective temperature versus gravity is close to but not
exactly a power law.

Observational data often show the polar-equator contrast in effective temperature
in terms of the exponent β defined as

$$T_{\text{eff}} \propto g_{\text{eff}}^{\beta} \tag{5.25}$$

We shall call this approximate modeling the "β-model". Actually, note that (5.25)
demands that

$$\beta = \frac{\partial \ln T_{\text{eff}}}{\partial \ln g_{\text{eff}}}\Big|_{r=R(\theta)} \tag{5.26}$$

where $R(\theta)$ is the radius of the star at colatitude θ. Since the relation between T_{eff}
and g_{eff} is not a power law, β is not constant on the surface of a rotating star. It varies
between two extreme values that we can also compute.

To make things simpler we therefore define the b-exponent as follows:

$$T_e = T_p \left(\frac{g_e}{g_p}\right)^b \qquad \text{or} \qquad b = \frac{\ln(T_e/T_p)}{\ln(g_e/g_p)} \tag{5.27}$$

where the indices e and p refer to the equator and pole respectively. T and g
designate the effective temperature and effective surface gravity.

From the polar and equatorial expression of the flux, we get

$$F_e = (1 - \omega^2)^{-2/3} g_e \qquad \text{and} \qquad F_p = e^{2\omega^2 r_p^3/3} g_p$$

for the ω-model, while, from the Roche model,

$$\frac{g_e}{g_p} = r_p^2(1 - \omega^2) \quad \text{with} \quad r_p = \frac{1}{1 + \omega^2/2}$$

where r_p is the polar radius. So we find

$$\left(\frac{T_e}{T_p}\right)^4 = \frac{(1 - \omega^2)^{1/3}}{(1 + \omega^2/2)^2} e^{-2\omega^2 r_p^3/3}$$

and

$$b = \frac{1}{4} - \frac{1}{6} \frac{\ln(1 - \omega^2) + \omega^2 r_p^3}{\ln(1 - \omega^2) - 2\ln(1 + \omega^2/2)} \tag{5.28}$$

We plotted in Fig. 5.5 the values of b with increasing values of the flatness (namely with increasing rotation). In this figure we see that the b-exponent is close to a linear dependence $b = \frac{1}{4} - \frac{1}{3}\varepsilon$ up to $\varepsilon = 0.3$. However, note that since the true dependence is not a power law, β, as given by (5.26), varies at the surface of a given star. We also show in Fig. 5.5 its range of variation and it is clearly not negligible when ε is larger than ~ 0.15. This means that if we had access to a very high spatial resolution of the stellar surface we would find different β's whether we look at the pole (large values) or at the equator (low values).

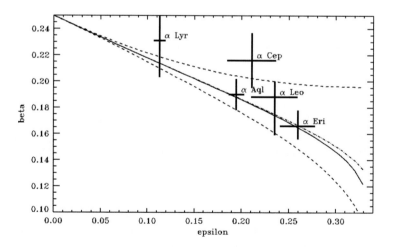

Fig. 5.5 The β-values from various models: The *solid line* shows the b-exponent of the ω-model, while the *dot-dashed line* shows the corresponding ESTER model. The extra *dashed lines* give the range of β values spawned at the stellar surface by a β-model. Data from interferometric observations of some early-type stars are shown (from Domiciano de Souza et al. 2014)

If the β-model is a poor representation of the latitudinal variation of the flux, can we devise a better one? Surely, a decomposition of the effective temperature on the spherical harmonics basis, namely

$$T_e(\theta) = \sum_{l,m} t_m^l Y_\ell^m$$

has the advantage of being model independent. The coefficients of the expansion are the results of observations. Such an expansion is already used for the description of spotted stars for the reconstruction of their magnetic fields (e.g. Donati et al. 2006 but see also the lecture of Kochukhov in this volume).

In Fig. 5.5, we also show the observationally derived values for a few early-type stars observed with interferometers. The matching is quite remarkable, even if some cases like α Cep certainly need a more detailed study.

To finish with the case of early-type stars, let us consider the case of small rotations. We may first derive the linear dependence of the b-exponent with ε. From (5.28) we get

$$b = \frac{1}{4} - \frac{1}{6}\omega^2 + \mathcal{O}(\omega^4) \qquad \text{or} \qquad b = \frac{1}{4} - \frac{1}{3}\varepsilon + \mathcal{O}(\varepsilon^2) \tag{5.29}$$

where we observed that $\varepsilon = 1 - r_p$. This expression shows that in the limit of small rotation we recover von Zeipel law. To understand the origin of this property, it is useful to reconsider the ω-model and the expression of the function $F(r, \theta)$. Let us first solve (5.18) in the limit $\omega \ll 1$. This yields

$$\psi = \theta + \frac{1}{3}\omega^2 \sin\theta \cos\theta + \mathcal{O}(\omega^4) \tag{5.30}$$

From this relation, we derive the asymptotic expression of $F(r, \theta)$ at low ω, namely

$$F(r, \theta) = 1 + \frac{2}{3}\omega^2 r^3$$

The latitudinal dependence has disappeared. Hence the latitudinal variations of the flux are those of the effective gravity. Therefore, von Zeipel law applies at low rotation rates. We can understand this result, if we recall that in the limit of zero rotation, the star is spherical and all surfaces of constant pressure, temperature, etc. are spheres so that we can consider the gravitational potential or the pressure as the independent variable. Thus we recover a kind of barotropic situation where one can use a relation between pressure and density, and derive a von Zeipel law.

5.3   The Case of Convective Envelopes

5.3.1   Lucy's Problem

In the 1960s it was realized that gravity darkening was very important for the interpretation of light curves of contact binaries (like the W UMa-type stars). But most of these stars are low-mass stars, thus with a convective envelope. The use of von Zeipel law, which is based on heat diffusion, was therefore doubtful.

So Lucy (1967) asked: "What is the gravity-darkening law appropriate for late-type stars whose subphotospheric layers are convective?" Lucy's reasoning was the following.

In the convective envelope of a rotating star, if we go deep enough, we should reach a medium of constant entropy. This value should be the same whatever the latitude. 1D models show that the entropy jumps from a minimum near the surface (where the convective driving ceases) to a plateau in the deep layers where convective mixing is efficient (see Fig. 5.6). Lucy argues that the value of the entropy s on this plateau is a function of the surface gravity g_s and effective temperature T_{eff}. He thus writes

$$s \equiv s(g_s, T_e) \tag{5.31}$$

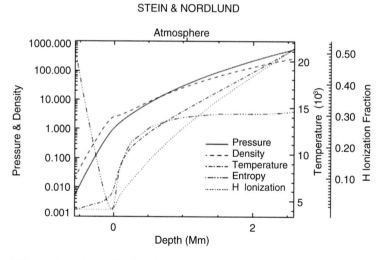

Fig. 5.6 Thermodynamic profile of the Sun according to Stein and Nordlund 1998

In the case of a rotating star, where g_s and T_{eff} vary, we must have

$$s(g_s, T_e) = s_0 \tag{5.32}$$

where s_0 is the entropy on the plateau. If we differentiate this expression with respect to g_s and T_e, we find that

$$\frac{\partial s}{\partial g_s} dg_s + \frac{\partial s}{\partial T_e} dT_e = 0$$

in the deep layers of the rotating star. Since we admit that $T_e \propto g_s^\beta$, then we have

$$\frac{\partial s}{\partial \ln g_s} + \beta \frac{\partial s}{\partial \ln T_e} = 0 \tag{5.33}$$

Thus, if we are able to evaluate the values of the above partial derivatives, we can obtain β. For that, Lucy considered various 1D neighbouring models (we do not know how the variations were made), and evaluated the partial derivatives so as to find β. Using five stellar models (three with $M = 1\,M_\odot$, two with $M = 1.26\,M_\odot$), he found that

$$0.069 \le \beta \le 0.088$$

Lucy adopted $\beta = 0.08$ as a representative value.

5.3.2 A New Derivation of Lucy's Result

It is interesting to note that Lucy's results may be derived from simple considerations on one dimensional stellar models in the solar mass range.

Let us first recall that the surface of a star is usually determined by a surface pressure given by

$$P = \frac{2g_s}{3\kappa} \tag{5.34}$$

where g_s is the surface gravity and κ an average opacity. This boundary condition comes from the assumption of hydrostatic equilibrium of the atmosphere, namely

$$\frac{\partial P}{\partial z} = -\rho g \quad \Longleftrightarrow \quad \frac{1}{\rho \kappa} \frac{\partial P}{\partial z} = -\frac{g}{\kappa} \quad \Longleftrightarrow \quad \frac{\partial P}{\partial \tau} = \frac{g}{\kappa}$$

where the last relation is integrated from the zero optical depth down to $\tau = 2/3$. In the range of density and temperatures typical of the solar type stars, opacity may be approximated by a power law of the form:

$$\kappa = \kappa_0 \rho^\mu T^{-s} \tag{5.35}$$

For instance Christensen-Dalsgaard uses $\mu = 0.408$ and $s = -9.283$ for the Sun (e.g. Christensen-Dalsgaard and Reiter 1995).

Now in convective envelopes, the variation of pressure and density are related to temperature through

$$P \propto T^{n+1} \quad \text{and} \quad \rho \propto T^n .$$

namely with a polytropic law with $n = 3/2$.

Using the foregoing power laws for the opacity, pressure and density, we can express gravity as a function of temperature. We find that

$$g \propto T^{n(\mu+1)+1-s}$$

Identifying temperature and effective temperature, we find a gravity darkening exponent which reads:

$$\beta = \frac{1}{n(\mu + 1) + 1 - s} \tag{5.36}$$

Using Christensen-Dalsgaard's solar values and $n = 3/2$, the foregoing expression yields

$$\beta \simeq 0.0807$$

which is precisely the value found by Lucy. This is no surprise since Lucy used models similar to solar models, so the power law fit of Christensen-Dalsgaard is appropriate.

This derivation clearly shows that this β-exponent, as defined by (5.33), depends on the chemical properties of the surface through the opacities.

5.3.3   Can Lucy's Law Represent a Gravity Darkening Effect?

The foregoing derivation of Lucy's result enlights us on the origin of Lucy's value of the β exponent. We see that it is essentially due to the strong dependence of opacity with temperature in the surface layers. Since the values for μ and s are chosen to fit the table values in some range of density and temperature, we understand that Lucy's result applies only to stars similar to the Sun, in terms of gravity and effective temperature. We may note, as Espinosa Lara and Rieutord (2012), that if the opacity law extend in the deep layers so as to control the structure of the envelope and leads to a radiative one, then $\beta = 1/4$ because the polytropic index is $n = (s + 3)/(\mu + 1)$. We recover the previous result for non rotating radiative envelopes. We see that when the opacity is such that the polytropic index is less than 3/2, and the envelope is convective, the β is governed by the opacity of the surface layers, those which are assumed to be transparent and fixing the atmosphere. The structure of the envelope is close to the adiabatic index n = 3/2.

Now we wish considering the case of rotating stars. The question is whether we can use the foregoing value of the exponent, if we consider a fast rotating star of solar type. A first obstacle is the validity of the boundary condition (5.34), which relies on a hydrostatic equilibrium. When rotation is present such an equilibrium is impossible because of baroclinicity (for the same reason as the origin of the so-called von Zeipel paradox, see Rieutord 2006). The proper boundary condition, replacing (5.34) should be derived from

$$\boldsymbol{v} \cdot \nabla \boldsymbol{v} = -\frac{1}{\rho} \nabla P - \nabla \Phi$$

where $\boldsymbol{v}$ is the fluid velocity in an inertial frame. Basically the flow is a differential rotation plus some weak meridional currents. The important point is that the differential rotation is latitude dependent. Hence, if we were to use some pressure boundary condition like (5.34), we should expect some extra variations from this latitudinal differential rotation. But this is likely not the whole story as we shall discuss it now.

If we consider the deep convective envelope of a rapidly rotating star, we might consider too contradicting effects. First the Coriolis effect: analysis of a linear stability of a convectively unstable layer shows that polar regions are less unstable than equatorial ones. This a consequence of the presence of the Coriolis force. This force indeed prevents variations of the velocity field along the rotation axis (the so-called Taylor–Proudman theorem). It shows up in numerical simulations as convective rolls parallel to the rotation axis near the equatorial regions (Glatzmaier and Olson 1993). For stars this may imply that the convective flux is larger in the equatorial plane than in the polar region, thus meaning a negative β. However, in the equatorial plane the effective gravity is less, and so is the buoyancy force. This is the effect of centrifugal force, which therefore points to more flux in the polar region (thus for a positive β). The conclusion of the foregoing argument is that nothing is clear. We may only guess that if Lucy's law applies, it is for slowly rotating stars of solar type only. This is a deceptive conclusion since we may have interesting data only on fast rotators or evolved stars with weak self-gravity. In addition, the previous remarks do not mention the magnetic fields that are almost unavoidable in late-type stars.

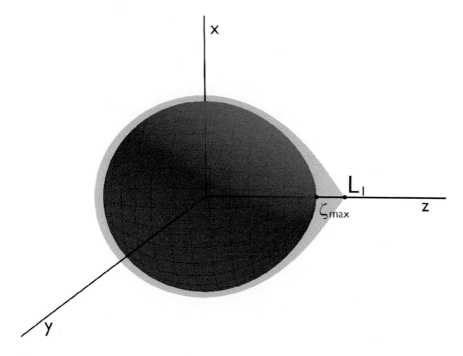

Fig. 5.7 Schematic representation of the primary star with filling factor $\rho = 0.8$. The position of the Lagrange point L_1 is shown on the z-axis that joins the centre of the two stars (from Espinosa Lara and Rieutord 2012)

5.4 Binary Stars

Binary stars is another domain where gravity darkening has been considered, mainly for reproducing the light curves of eclipsing binaries. We may wonder if the ω-model can be generalized to predict the gravity darkening of a star belonging to a binary system. It does but without any (known) analytic solution.

Let us follow the work of Espinosa Lara and Rieutord (2012). In the radiative envelope of an early-type star member of a binary system we can still write the conservation of the flux and assume the anti-parallelism of flux and effective gravity:

$$\mathrm{Div}\boldsymbol{F} = 0 \qquad \text{and} \qquad \boldsymbol{F} = -f\boldsymbol{g}_{\mathrm{eff}}$$

but now the effective gravity comes from the 3D potential:

$$\phi = -\frac{GM_1}{r} - \frac{GM_2}{\sqrt{a^2 + r^2 - 2ar\cos\theta}} - \frac{1}{2}\Omega^2 r^2(\sin^2\theta\sin^2\varphi + \cos^2\theta) + a\frac{M_2}{M_1 + M_2}\Omega^2 r\cos\theta , \tag{5.37}$$

where M_1 and M_2 are the masses of the two stars, 'a' is the distance between the two stellar centres and Ω is the orbital angular velocity. The orbit is assumed circular.

Let us write $\mathrm{Div}(f\boldsymbol{g}_{\mathrm{eff}}) = 0$ as

$$\boldsymbol{n} \cdot \nabla \ln f = \frac{\nabla \cdot \boldsymbol{g}_{\mathrm{eff}}}{g_{\mathrm{eff}}} , \qquad (5.38)$$

where we set $\boldsymbol{g}_{\mathrm{eff}} = -g_{\mathrm{eff}}\boldsymbol{n}$.

We consider the three-dimensional curve $\mathcal{C}(\theta_0, \varphi_0)$ that starts at the centre of the star with the initial direction given by (θ_0, φ_0), and that is tangent to $\boldsymbol{n}$ at every point. $\mathcal{C}(\theta_0, \varphi_0)$ is therefore a field line of the effective gravity field.

The value of f at a point $\boldsymbol{r}$ along the curve can be calculated as a line integral

$$f(\boldsymbol{r}) = f_0 \exp\left(\int_{\mathcal{C}(\theta_0,\varphi_0)} \frac{\nabla \cdot \boldsymbol{g}_{\mathrm{eff}}}{g_{\mathrm{eff}}} \, \mathrm{d}l\right) \qquad \text{for } \boldsymbol{r} \in \mathcal{C}(\theta_0, \varphi_0) . \qquad (5.39)$$

Despite much efforts no analytical expression could be found for f. Expression (5.39) is thus integrated numerically.

One interesting result of this approach, is that there is not a one-to-one relation between effective gravity and effective temperature. Indeed, because of the absence of symmetry of the star (except of the equatorial one if the obliquity is zero), two different points of the stellar surface may have the same effective gravity but a different effective temperature. This property comes from expression (5.39): the path integrals that lead to two points of identical effective gravity are not necessarily the same and can lead to different values of the flux. This property is illustrated in Fig. 5.8. In this figure we see that the curve $T_{\mathrm{eff}} = f(g_{\mathrm{eff}})$ is not smooth because

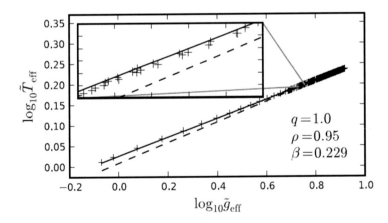

Fig. 5.8 From Espinosa Lara and Rieutord (2012), correlation of effective temperature and effective gravity in the primary early-type star of a binary system. Mass ratio is unity and the star fills the Roche lobe at 95 % (see text for our definition). The correlation may be represented by a β-exponent of 0.229. The *solid line* shows a linear fit, the *dashed line* the von Zeipel law, and pluses are from our generalized ω-model

similar values of g_{eff} lead to different values of T_{eff}. Fortunately, these variations are small.

As in Espinosa Lara and Rieutord (2012), we define q as the mass ratio, and evaluate the filling of the Roche lobe by the radius ρ of the star along the line joining the stellar centres, taking the distance between the star centre and the Lagrange L_1 point as unity. Hence, a star filling its Roche lobe has $\rho = 1$ while the one filling it at 95 % has $\rho = 0.95$ (see Fig. 5.7). The different positions where the same effective temperature are found, is illustrated in the two cases shown in Fig. 5.9. There we see that the curves of isoflux are not simple curves over the stellar surface.

As shown by Djurašević et al. (2003), the determination of the β-exponent from the light curves of semi-detached binaries is almost impossible since magnetic spots induce similar variations (see Figs. 5.10 and 5.11).

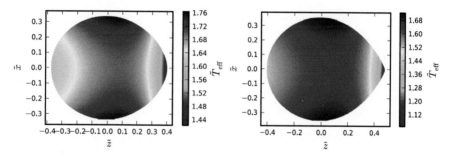

Fig. 5.9 From Espinosa Lara and Rieutord (2012): distribution of the effective temperature at the surface of a tidally distorted star. *Left*: $q = 1$ and $\rho = 0.8$. *Right*: $q = 1$ and $\rho = 0.95$

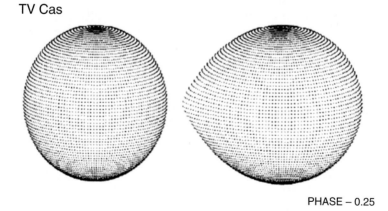

Fig. 5.10 A model of TV Cas that leads to $\beta = 0.15$ from fitting the light curve (from Djurašević et al. 2003)

TV Cas

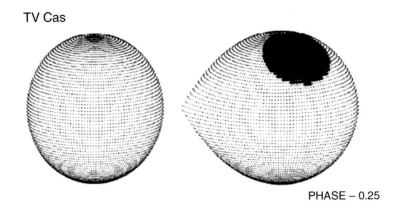

PHASE – 0.25

Fig. 5.11 The second model with a spot and $\beta = 0.25$ for TV Cas (from Djurašević et al. 2003); the difference between the calculated and observed light curve is the same as with the model of Fig. 5.11

5.5 Conclusions

To conclude these notes, I would like to stress a few points on gravity darkening:

- As far as non-magnetic early-type stars are concerned, gravity darkening has no longer to be proved. The use of the β-model, which is not physically sound can be left aside and replaced by the ω-model, which has the advantage of giving a direct estimate of the ω parameter.
- As far as late-type stars or giant stars are concerned, the situation is much more uncertain. The problem is indeed more difficult both on the theoretical and observational sides. On the theoretical side, the absence of any universally accepted model of turbulent rotating convection impedes any serious prediction on the latitude dependence of the convective flux. Observational constraints are therefore most welcome. However, this is not a simple matter either. Convective envelopes are usually harboring magnetic fields which can disturb the flux distribution. Ideally, the surface of these stars should be constrained by both interferometers and Zeeman-Doppler Imaging so as to disentangle the effects.
- Finally, for both type of (single) stars, we may recommend the following scheme of hypothesis and measurements. First assume the axi- and equatorial symmetry of the star. Then, if the star is centrally condensed (like an early-type or a giant one), adopt the Roche model. If the star is not centrally condensed (like a late-type star of the main sequence), a bipolytropic model is fine. Such a model, which fits the radiative and convective zones with a polytrope, just depends on three parameters, mass, equatorial radius and ω, just like the Roche model. Then, the flux or the effective temperature $T_{\mathrm{eff}}(\theta)$ can be derived after an expansion on the spherical harmonic basis along with an atmosphere model used for the

determination of the limb darkening effect. The gravity darkening law can then be evaluated from the curve (or correlation) $T_{\mathrm{eff}}(\theta)$ versus $g_{\mathrm{eff}}(\theta)$.

Acknowledgements I am grateful to the organizers of the Besançon school for their invitation, and the opportunity to present in more details the recent work I did with Francisco Espinosa Lara on gravity darkening. This school triggered many stimulating discussions that helped me deepen this subject. Finally, I would like to stress that this work owes much to Francisco who had the original idea of the ω-model.

Appendix: Angular Velocity with Respect to Critical Rotation

In this appendix we discuss the correspondence between two definitions of the scaled angular velocity. The first one is the one we used in the text, namely

$$\omega = \frac{\Omega}{\Omega_k} = \Omega \left(\sqrt{\frac{GM}{R_e^3}} \right)^{-1}$$

where Ω_k is the orbital angular velocity for an orbit at the actual equatorial radius of the star.

The other definition is based on the Roche model and considers the angular velocity Ω_c such that the rotation on the equatorial radius is keplerian. This latter definition is a true critical angular velocity, while the previous one is a keplerian velocity at the actual equatorial radius. However, the critical angular velocity is model dependent, this is why we have to mention Roche's model for the definition of Ω_c. The first definition does not need any model, but it is not the exact critical angular velocity. This latter quantity cannot in general be computed a priori with just a given spherical model of a star. It needs a full computation of the structure at the actual critical velocity and is thus an output of 2D models like ESTER ones (e.g. Espinosa Lara and Rieutord 2013).

So we now only consider Roche models where all quantities can be derived in a simple manner. We recall that the polar R_p and equatorial R_e of an equipotential of a star rotating at angular velocity Ω are related by

$$\frac{GM}{R_p} = \frac{GM}{R_e} + \frac{1}{2}\Omega^2 R_e^2 \tag{5.40}$$

Then, the critical angular velocity Ω_c and the critical equatorial radius R_{ec} are related by

$$\Omega_c^2 = \frac{GM}{R_{ec}^3} \tag{5.41}$$

and hence

$$R_{ec} = \frac{3}{2}R_p \tag{5.42}$$

at critical rotation.

If the rotation is subcritical, the Roche model gives the following relation between R_e and R_p

$$R_p = R_e \left(1 + \frac{\omega^2}{2}\right)^{-1} \tag{5.43}$$

But we may take Ω_c as the scale of the rotation rate and set

$$\tilde{\omega} = \frac{\Omega}{\Omega_c} \tag{5.44}$$

From the preceding definitions we get the relation between $\tilde{\omega}$ and ω, namely

$$\tilde{\omega} = \omega\sqrt{\frac{27}{8}} \left(1 + \omega^2/2\right)^{-3/2} \tag{5.45}$$

We note that if $\omega = 1$ then $\tilde{\omega} = 1$ as expected. We also note that if Ω is subcritical, then $R_e < 3R_p/2$ and therefore $\Omega_k > \Omega_c$, which implies that we always have

$$\tilde{\omega} \geq \omega \tag{5.46}$$

Equation (5.45) shows that it is easy to compute $\tilde{\omega}$ from ω but the opposite is a little more complicated since a cubic equation must be solved. Setting $\chi = \arcsin \tilde{\omega}$, we find

$$\omega = \sqrt{\frac{6}{\tilde{\omega}} \sin(\chi/3) - 2} \tag{5.47}$$

References

Alecian, G. (2013). Atomic diffusion in the atmospheres of upper main-sequence stars. In *New advances in stellar physics: From microscopic to macroscopic processes. EAS Publications Series* (Vol. 63, pp. 219–226).

Christensen-Dalsgaard, J., & Reiter, J. (1995). A comparison of precise solar models with simplified physics. In R. K. Ulrich, E. J. Rhodes, Jr., & W. Dappen (Eds.), *ASP Conf. Ser. 76: GONG 1994. Helio- and Astro-Seismology from the Earth and Space* (p. 136).

Djurašević, G., Rovithis-Livaniou, H., Rovithis, P., Georgiades, N., Erkapić, S., & Pavlović, R. (2003). Gravity-darkening exponents in semi-detached binary systems from their photometric observations: Part I. *Astronomy and Astrophysics, 402*, 667–682.

Domiciano de Souza, A., Kervella, P., Moser Faes, D., Dalla Vedova, G., Mérand, A., Le Bouquin, J.-B., et al. (2014). The environment of the fast rotating star Achernar. III. Photospheric parameters revealed by the VLTI. *Astronomy and Astrophysics, 569*, A10.

Donati, J.-F., Howarth, I. D., Jardine, M. M., Petit, P., Catala, C., Landstreet, J. D., et al. (2006). The surprising magnetic topology of τ Sco: Fossil remnant or dynamo output? *Monthly Notices of the Royal Astronomical Society, 370*, 629–644.

Espinosa Lara, F., & Rieutord, M. (2011). Gravity darkening in rotating stars. *Astronomy and Astrophysics, 533*, A43.

Espinosa Lara, F., & Rieutord, M. (2012) Gravity darkening in binary stars. *Astronomy and Astrophysics, 547*, A32.

Espinosa Lara, F., & Rieutord, M. (2013). Self-consistent 2D models of fast rotating early-type stars. *Astronomy and Astrophysics, 552*, A35.

Glatzmaier, G., & Olson, P. (1993) Highly supercritical thermal convection in a rotating spherical shell: Centrifugal vs. radial gravity. *Geophysical and Astrophysical Fluid Dynamics, 70*, 113–136.

Korhonen, H., González, J. F., Briquet, M., Flores Soriano, M., Hubrig, S., Savanov, I., et al. (2013). Chemical surface inhomogeneities in late B-type stars with Hg and Mn peculiarity. I. Spot evolution in HD 11753 on short and long time scales. *Astronomy and Astrophysics, 553*, A27.

Lucy, L. B. (1967). Gravity-darkening for stars with convective envelopes. *Zeitschrift für Astrophysik, 65*, 89.

Monnier, J., Che, X., Zhao, M., Ekström, S., Maestro, V., Aufdenberg, J., et al. (2012). Resolving vega and inclination controversy with CHARA/MIRC. *The Astrophysical Journal, 761*, L3.

Rieutord, M. (2006). On the dynamics of radiative zones in rotating star. In M. Rieutord, & B. Dubrulle (Eds.), *Stellar fluid dynamics and numerical simulations: From the sun to neutron stars. EAS* (Vol. 21, pp. 275–295).

Rieutord, M., & Rincon, F. (2010). The sun's supergranulation. *Living Reviews in Solar Physics, 7*, 1–70.

Stein, R. F., & Nordlund, Å. (1998). Simulations of solar granulation. I. General properties. *The Astrophysical Journal, 499*, 914.

Vauclair, S., & Vauclair, G. (1982). Element segregation in stellar outer layers. *Annual Review of Astronomy and Astrophysics, 20*, 37–60.

von Zeipel, H. (1924). Zum Strahlungsgleichwicht der Sterne. In *Probleme der Astronomie (Festschrift für H. von Seeliger)* (pp. 144–152). Heidelberg: Springer.

Chapter 6
Interferometry to Determine Stellar Shapes: Application to Achernar

Pierre Kervella

Abstract The shape of stellar photospheres can depart significantly from the spherical geometry, due e.g. to fast rotation. In this chapter, I focus on the application of long-baseline interferometry to the determination of the photospheric shape of fast rotating stars. I present the example of the VLT Interferometer observations of the nearby Be star Achernar (α Eri), using the VINCI (two telescopes) and PIONIER (four telescopes) beam combiners. I present the adjustment of a simplified model of the light distribution of Achernar to the measured interferometric visibilities and closure phases. This example application is based on the LITpro software from the JMMC.

6.1 Introduction

Optical interferometry at visible and infrared wavelengths over hectometric baselines provides a sufficiently high angular resolution to spatially resolve the photosphere of nearby stars. This gives the possibility to measure their size, shape and surface brightness distribution, that are intimately linked to their physical properties (internal structure, rotation, convection, . . .).

The southern star Achernar (α Eridani, HD 10144) is the brightest and one of the nearest Be stars ($m_V = 0.46$, $\pi = 22.7 \pm 0.6$ mas). Its estimated projected rotation velocity $v \sin i$ ranges from 220 to 270 km/s and the effective temperature T_{eff} is around 15,000 K (Domiciano de Souza et al. 2014). Such rapid rotation ($\approx 80\,\%$ of the critical velocity) induces two effects on the star structure: a rotational flattening and an equatorial darkening (von Zeipel 1924).

The extreme flattening of the photosphere of Achernar was first measured by Domiciano de Souza et al. (2003). This discovery turned Achernar into a prominent

P. Kervella (✉)
LESIA, Observatoire de Paris, France

Unidad Mixta Internacional Franco-Chilena de Astronomía (UMI 3386), CNRS/INSU, Paris, France

Departamento de Astronomía, Universidad de Chile, Santiago, Chile
e-mail: pierre.kervella@obspm.fr

© Springer International Publishing Switzerland 2016   127
J.-P. Rozelot, C. Neiner (eds.), *Cartography of the Sun and the Stars*,
Lecture Notes in Physics 914, DOI 10.1007/978-3-319-24151-7_6

fiducial object to study the effect of rotation on the structure and atmospheric properties of fast rotating stars. A recent review of the field of optical interferometry applied to fast rotators can be found in van Belle (2012) and observation reports in Aufdenberg et al. (2006) and Monnier et al. (2014). As a Be star, Achernar presents episodic emission lines in its spectrum, created by the temporary appearance of an equatorial disk. Due to the high temperature of its polar caps, is shows a relatively strong polar wind (Kervella and Domiciano de Souza 2006). It was also discovered recently by Kervella and Domiciano de Souza (2007, see also Kervella et al. 2008) that Achernar is a binary star with a close main sequence companion of early A spectral type.

6.2   VLTI Observations of Achernar

Thanks to its brightness and southern declination ($\delta = -57°$), Achernar has been observed regularly since the first light of the VLT Interferometer in 2001 (Glindemann et al. 2003; Mérand et al. 2014), first using the VINCI instrument (Domiciano de Souza et al. 2003; Kervella and Domiciano de Souza 2006), then with the beam combiners MIDI (Kervella et al. 2009), AMBER (Domiciano de Souza et al. 2012) and PIONIER (Domiciano de Souza et al. 2014).

The observations discussed here were obtained with the two-telescope beam combiner VINCI (Kervella et al. 2004) in the H (1.64 μm) and K (2.2 μm) bands, and the four-telescope instrument PIONIER (Le Bouquin et al. 2011) in the H band. The details of the observations with these two instruments are presented respectively in Kervella and Domiciano de Souza (2006) and Domiciano de Souza et al. (2014). Photographs of the different telescopes used for these observations and of one of the long optical delay lines of the VLTI are shown in Fig. 6.1.

The VINCI observations produced a set of squared interferometric visibility measurements over a range of azimuth angles and projected baseline lengths. Thanks to the larger number of telescopes, the observations obtained with PIONIER

Fig. 6.1 *Left:* Light collectors used for the observations of Achernar: 0.4 m test siderostat (*foreground*), 1.8 m auxiliary telescopes (*right*) and 8.2 m unit telescopes (*background*). *Right:* One of the long delay lines of the VLTI

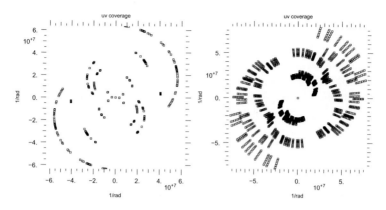

Fig. 6.2 Coverage of the (u, v) plane of the spatial frequencies for the VINCI (*left*) and PIONIER (*right*) observations of Achernar

resulted in a much denser coverage of the (u, v) plane of the spatial frequencies (Fig. 6.2). The PIONIER spectral dispersion over seven channels results in short segments in the (u, v) plane, as opposed to the single points sampled by the VINCI instrument. In addition, the six baselines sampled simultaneously by PIONIER give access to three independent closure phase triangles. The closure phase is a particularly interesting observable for fast rotating stars as it is related to the degree of deviation of the observed object from point symmetry. As the polar caps of fast rotators are brighter than their equatorial belt (due to the Von Zeipel effect), the closure phase is an excellent proxy for the inclination of the rotation axis on the line of sight.

6.3 Analysis Using LITpro

The objective of the application presented here is to determine the equatorial angular size, flattening ratio and orientation on sky of Achernar's apparent disk. With the PIONIER closure phases, we will also probe the degree of asymmetry of the flux distribution at the surface. The VINCI data were processed using the dedicated pipeline,[1] and then converted to the OIFITS data format. As discussed by Domiciano de Souza et al. (2014), the raw data from the PIONIER instrument have been reduced with the dedicated pipeline[2] available from IPAG in France, that produced calibrated squared visibilities and phase closures.

---

[1] *vndrs* package, see Kervella et al. (2004).

[2] *pndrs* package, see Le Bouquin et al. (2011).

As a practical example of the analysis of interferometric data using a model-fitting approach, I present an analysis of the OIFITS files using the `LITPro`[3] software (Tallon-Bosc et al. 2008). For simplicity, I consider a uniform ellipse as the morphological model to adjust to the VINCI and PIONIER data. This model has three parameters in total: the major axis of the ellipse a (in milliarcseconds, hereafter mas), the flattening ratio $f = a/b$ (with b the minor axis) and the position angle of the minor axis θ_{pol} (corresponding to the polar axis position angle with respect to North, in degrees).

It should be noted that the uniform ellipse model is only a poor match for the actual intensity distribution of Achernar. As shown by Domiciano de Souza et al. (2014), the polar temperature reaches more than 17,000 K, while the equatorial temperature is below 13,000 K. This results in a strongly non-uniform brightness of its apparent disk, as the polar caps have a surface brightness three times higher than the equator. Moreover, as shown by Kervella and Domiciano de Souza (2006), Achernar is surrounded by an extended envelope, that we neglect in this simplified approach. We also neglect the contribution of the close-in stellar companion of Achernar, that amounts to only ≈3 % of the primary flux in the near-infrared.

6.3.1 VINCI Data Set Only

We first fit the VINCI squared visibilities alone, in order to be able to compare with the results of a global fit including the extensive PIONIER data set. The results are presented in Fig. 6.3, and the derived parameters are listed in Table 6.1.

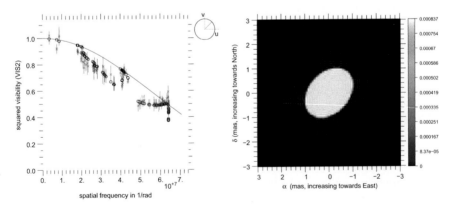

Fig. 6.3 *Left:* VINCI squared visibilities of Achernar (*red crosses* with error bars), uniform ellipse model visibilities (*black circles*) and polar visibility curve of the model (*blue curve*). *Right:* Image of the best-fit uniform ellipse model

---

[3]LITpro software, available at http://www.jmmc.fr/litpro.

Table 6.1 Parameters of the uniform ellipse model of Achernar

Data set	$f = b/a$	a (mas)	θ_{pol} (deg)	χ^2_{red}
VINCI	1.39 ± 0.04	2.33 ± 0.06	45.6 ± 1.9	2.8
VINCI+PIONIER	1.239 ± 0.005	1.914 ± 0.005	36.1 ± 0.7	2.3

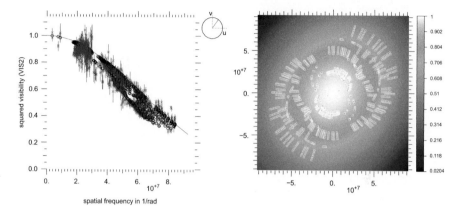

Fig. 6.4 *Left:* PIONIER squared visibilities of Achernar and adjusted model (see Fig. 6.3 for symbols). *Right:* Two dimensional visibility function of the best-fit uniform ellipse model (color scale) and position of the sampled (u, v) points

The reduced χ^2 of the fit of 2.8 shows that the chosen model is not a very good representation of the data. This is particularly visible in Fig. 6.3 for the lower spatial frequencies, for which the observed visibilities are systematically below the model visibilities. This behavior is caused by the presence of spatially extended emission. This envelope was identified by Kervella and Domiciano de Souza (2006) using a model combining a uniform ellipse and a Gaussian extended component.

6.3.2 Full VINCI+PIONIER Data Set

The (u, v) plane coverage of the VINCI+PIONIER data sets is much denser and more uniform than the VINCI data alone, as shown in Fig. 6.2. The best-fit parameters of the uniform ellipse model are listed in Table 6.1. These parameters are in reasonably good agreement with those derived by Domiciano de Souza et al. (2014) using a physically realistic model of the star. The reduced χ^2 of the fit is slightly better than for the VINCI data alone, although it still shows that our simple ellipse model is insufficient to interpret the interferometric data (Fig. 6.4).

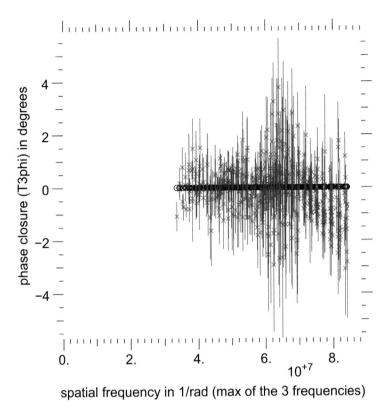

Fig. 6.5 Closure phases measured by PIONIER on Achernar (*red crosses* with error bars). The centrally symmetric uniform ellipse model (*black circles*) has naturally a zero closure phase at all spatial frequencies

Figure 6.5 shows the closure phase signal from the PIONIER observations. As the uniform ellipse model is point-symmetric, the corresponding closure phases are always equal to zero. This corresponds well to the observed values, indicating that the actual light distribution on Achernar is close to centrally symmetric. Taking into account the Von Zeipel effect, the inclination of the rotation axis on the line of sight can therefore be either close to $0°$ (i.e. the star is seen pole-on) or $90°$ (the star is seen equator-on). Considering that the measured flattening is large, this implies that the star is seen close to equator-on. An almost pole-on star would appear circular, as in the case of Vega (Aufdenberg et al. 2006). This is qualitatively consistent with the results obtained recently by Domiciano de Souza et al. (2014) for Achernar, who determined a high inclination of $\approx 61°$.

6.3.3   Discussion

The combined analysis of the VINCI+PIONIER results in significantly different
model parameters than with the VINCI data alone. Figure 6.6 shows a graphical
view of the two resulting ellipses. While the position angle of the polar axis on the
sky is reasonably similar, the flattening ratio is significantly lower for the global
fit than for the VINCI data only. This difference may be of astrophysical origin,
as Achernar is a Be star, and is therefore episodically surrounded by an equatorial
gaseous disk. The presence of such a disk at the epoch of the VINCI observations
and not at the epoch of the PIONIER observations would result in a larger apparent
equatorial size.

But more likely, the true reason for this difference is the limited (u, v) coverage
of the VINCI observations, compared to the PIONIER data. In particular, as shown
in Fig. 6.2, the position angle corresponding to the equator ($\approx$36°) is poorly covered
by the VINCI measurements, as only shorter baselines are available in this range of
position angle compared to the polar direction.

Finally, the model that is adjusted to the data (uniform ellipse) is an imperfect
match to the actual light distribution of Achernar. The true light distribution of
Achernar induces a baseline and azimuth dependence of the visibility that is not
reproduced properly by the model, and causes biases in the derived parameters.

Fig. 6.6 Best-fit model representations using the VINCI visibilities alone (larger ellipse), and the
full VINCI+PIONIER data set (smaller ellipse)

6.3.4 Conclusion

Using a very simple uniform ellipse model, we established that the derived flattening ratio $f = a/b$ of Achernar depends on the data set that is selected for the fit (VINCI alone or VINCI+PIONIER). This undesirable effect is caused by the relatively poor (u, v) coverage of our VINCI observations, and the fact that the uniform ellipse model is insufficient to reproduce properly the two dimensional visibility function of the star in the (u, v) plane. As the model visibility function is adjusted to a limited number of measurements, the fitted parameters depend on the distribution of the sampled spatial frequency in the data set.

To interpret properly the interferometric observations, the solution is to adopt a model of the stellar light distribution as realistic as possible. It can be assembled from a combination of the observed properties of the star (e.g. from spectroscopy, photometry, ...) and its expected internal structure from stellar interior models (Espinosa Lara and Rieutord 2007; Rieutord 2013). An example of a suitable model of Achernar is presented in Domiciano de Souza et al. (2014), where an application of image reconstruction to this star is also discussed. It is in any case essential to have a (u, v) plane coverage sufficiently dense and uniform (as the PIONIER observations) to prevent biases on the derived parameters (in particular the position angle of the polar axis on the sky).

Once these conditions are met, optical interferometry is extremely constraining to determine stellar shapes and other physical parameters linked to the intensity distribution on the photosphere. As interferometric measurements are independent from the other classical disk-integrated observables, the optimal approach is in most cases to include as much independent observables in the input data set (e.g. spectroscopy, photometry, ...) while keeping the number of free parameters in the model as low as possible.

Acknowledgements I am grateful to organizers of the Besançon school for their invitation. This research has made use of the Jean-Marie Mariotti Center LITpro service co-developed by CRAL, LAOG and FIZEAU. This research received the support of PHASE, the high angular resolution partnership between ONERA, Observatoire de Paris, CNRS and University Denis Diderot Paris 7. I acknowledge financial support from the "Programme National de Physique Stellaire" (PNPS) of CNRS/INSU, France. I used the SIMBAD and VIZIER databases at the CDS, Strasbourg (France), and NASA's Astrophysics Data System Bibliographic Services.

References

Aufdenberg, J. P., Mérand, A., Coudé du Foresto, V., et al. (2006). *The Astrophysical Journal, 645,* 664.

Domiciano de Souza, A., Hadjara, M., Vakili, F., et al. (2012). *Astronomy & Astrophysics, 545,* A130.

Domiciano de Souza, A., Kervella, P., Jankov, S., et al. (2003). *Astronomy & Astrophysics, 407,* L47.

Domiciano de Souza, A., Kervella, P., Moser Faes, D., et al. (2014). *Astronomy & Astrophysics, 569*, A10.

Espinosa Lara, F., & Rieutord, M. (2007). *Astronomy & Astrophysics, 470*, 1013.

Glindemann, A., Algomedo, J., Amestica, R., et al. (2003). Interferometry for optical astronomy II. In W. A. Traub (Ed.), *Society of Photo-Optical Instrumentation Engineers (SPIE) Conference Series*(Vol. 4838, pp. 89–100).

Kervella, P., & Domiciano de Souza, A. (2006). *Astronomy & Astrophysics, 453*, 1059.

Kervella, P., & Domiciano de Souza, A. (2007). *Astronomy & Astrophysics, 474*, L49.

Kervella, P., Domiciano de Souza, A., & Bendjoya, P. (2008). *Astronomy & Astrophysics, 484*, L13.

Kervella, P., Domiciano de Souza, A., Kanaan, S., et al. (2009). *Astronomy & Astrophysics, 493*, L53.

Kervella, P., Ségransan, D., & Coudé du Foresto, V. (2004). *Astronomy & Astrophysics, 425*, 1161.

Le Bouquin, J.-B., Berger, J.-P., Lazareff, B., et al. (2011). *Astronomy & Astrophysics, 535*, A67.

Mérand, A., Abuter, R., Aller-Carpentier, E., et al. (2014). *Proceedings of the SPIE*, (Vol. 9146, ID. 91460J).

Monnier, J. D., Che, X., Zhao, M., & ten Brummelaar, T. (2014). Resolving the future of astronomy with long-baseline interferometry. In M. J. Creech-Eakman, J. A. Guzik, & R. E. Stencel (Eds.), *Astronomical Society of the Pacific Conference Series* (Vol. 487, p. 137).

Rieutord, M. (2013). ESTER: Evolution STEllaire en Rotation, Astrophysics Source Code Library.

Tallon-Bosc, I., Tallon, M., Thiébaut, E., et al. (2008). *Society of Photo-Optical Instrumentation Engineers (SPIE) Conference Series* (Vol. 7013).

van Belle, G. T. (2012). *Astronomy & Astrophysics Review, 20*, 51.

von Zeipel, H. (1924). *Monthly Notices of the Royal Astronomical Society, 84*, 665.

Chapter 7
Interferometry to Image Surface Spots

Guy Perrin

Abstract I present in this lecture the technique of interferometric imaging at optical/infrared wavelengths. The technique has matured since the pioneering work of Michelson at the end of the XIXth—beginning of the XXth when he first resolved the surface of a star, Betelgeuse, with his colleague Pease. Images were obtained for the first time 20 years ago with the COAST instrument and interferometers have made constant progress to reach the minimum level where blind image reconstruction can be achieved. I briefly introduce the topic to recall why studying the surface and close environment of stars is important in some fields of stellar physics. I introduce the theory of imaging with telescopes and interferometers. I discuss the nature of interferometric data in this wavelength domain and give a few insights on the importance of getting access to visibility phases to obtain information on asymmetries of stellar surfaces. I then present the issue of aperture synthesis with a small number of telescopes, a signature of optical/infrared interferometers compared to the radio domain. Despite the impossibility to measure the phase of visibilities because of turbulence I show that useful information can be recovered from the closure phase. I eventually introduce the principles of image reconstruction and I discuss some recent results on several types of stars.

7.1 Motivations for Interferometric Imaging of Stars

The reasons to image the surface of stars to understand stellar physics are many whatever the evolution stage:

- Departure from spherical symmetry because of rotation or interactions in binaries;
- Intensity variations even in the simplest cases: basic limb darkening, gravitational darkening for rotating stars, convection, granulation;

G. Perrin (✉)
LESIA, Observatoire de Paris, PSL Research University, CNRS, Univ. Pierre et Marie Curie Paris 06, Sorbonne Université, Univ. Paris-Diderot, Sorbonne Paris-Cité, 5 place Jules Janssen, 92195 Meudon, France
e-mail: guy.perrin@obspm.fr

© Springer International Publishing Switzerland 2016   137
J.-P. Rozelot, C. Neiner (eds.), *Cartography of the Sun and the Stars*,
Lecture Notes in Physics 914, DOI 10.1007/978-3-319-24151-7_7

- The spatial variations of magnetic field and the link with convection in red supergiants for example;
- Spatial variations of surface brightness in the atmosphere of evolved stars;
- Interactions with the close circumstellar environment (e.g. understanding the onset of mass loss);
- Other reasons, the list could be quite long!

The problem of mass loss in red supergiants is a privileged science case for the author. The case for lower mass stars going through the asymptotic giant branch and in particular through the Mira phase is much better understood. The large amplitude pulsations input enough mechanical energy to lift the atmosphere up to an altitude where dust can condense. Radiation pressure from the star pushes the grains away to produce a wind in which the gas is dragged to reach high velocities of a few 10 km/s.

This scenario does not apply to red supergiants by lack of sufficient large amplitude pulsations. However, dust is detected at distances of a few tens of stellar radii from the photosphere of red supergiants (e.g. Danchi et al. 1994). Other sources of mechanical energy therefore need to be invoked. Betelgeuse has been a privileged playground for this quest because it is the largest red supergiant seen from Earth. One source of energy may be produced by large-scale convective motions who generate turbulent pressure which may be at the origin of mass loss by decreasing the effective gravity (Josselin and Plez 1997). Those upward and downward motions were detected in CO lines by Ohnaka et al. (2009) with the AMBER instrument at VLTI. Magnetic fields may be another source of energy. Haubois et al. (2009) have imaged convective cells at the surface of Betelgeuse with the IONIC instrument at IOTA and Aurière et al. (2010) have detected a 1 G magnetic field by Zeeman splitting with the NARVAL instrument at the Bernard Lyot Telescope at Pic du Midi. The existence of the latter may be caused by a dynamo effect in the convective cells. Convection is the key in this mechanism too. Given the large size of the convective cells and their characteristic time scales, one expects inhomogeneities in the atmospheric layers at the base of the mass loss process. Images taken by Kervella et al. (2009) with the NACO instrument at VLT show plumes that could be connected to the process of mass loss and that are consistent with these scenarios.

The understanding of the origin of mass loss in red supergiants probably comes with a complete picture of the zone located just above the photosphere where it starts from. The current largest telescopes can barely resolve the surface of the largest stars like Betelgeuse. The future Extremely Large Telescopes (ELTs) will allow to resolve the surface of the stars larger than 10 mas at near-infrared wavelengths. Very few stars match this requirement as only four red supergiants have sizes larger than 10 mas (counts base on the CHARM2 catalog by Richichi, Percheron & Khristoforova 2005). As a consequence, a much higher angular resolution is required. The milli-arcsecond scale is a good goal as it will allow to produce detailed images of the largest stars and corresponds to the resolution limit for a few tens of evolved stars at near-infrared wavelengths. Such an angular resolution can only be obtained by a large optical instrument with a diameter of 100–200 m depending on wavelength. Diluted apertures need to be put together to reach such scales. It is the goal of this lecture to explore how images can be obtained with this type of instrument.

7.2 Theory of Imaging with Telescopes and Interferometers

7.2.1 Imaging with Single Apertures

The theory of imaging of uncoherent sources in the most simple case of a linear system invariant by translation leads to the following relation between the spatial intensity distribution of the source and the image as a function of the angular coordinates α and β:

$$Im(\alpha, \beta) = O(\alpha, \beta) \star PSF(\alpha, \beta) \tag{7.1}$$

where PSF is the Point Spread Function, the image of a perfectly pointlike source. In the case where imaging is limited by the diffraction limit of an optical system, the PSF is the Fourier Transform of the autocorrelation function of the pupil:

$$PSF(\alpha, \beta) = \int\int P \otimes P(u, v)e^{-2i\pi(\alpha u + \beta v)}\,dudv \tag{7.2}$$

where the pupil function P is equal to 1 in the pupil and 0 outside, u and v are the spatial frequency coordinates and correspond to linear coordinates in the pupil plane divided by the wavelength. $P \otimes P(u, v)$ is the Optical Transfer Function of the imaging system. According to Eq. 7.1, it acts as a filter on the source spatial intensity distribution. For a circularly symmetric single aperture of diameter D, the OTF is a low-pass filter with cut-off frequency D/λ. All frequencies are set to 0 above this cut-off frequency by the imaging process. This sets the angular resolution to λ/D in absence of aberrations. All frequencies up to the cut-off frequency are present in the image, although they are all the more attenuated as they are close to the cut-off frequency. The image is not a perfect representation of the source but the presence of all frequencies up to the cut-off frequency makes it a useful image for astronomical interpretation, unless the size of the object is close to the diffraction limit and then a deconvolution process is necessary to disentangle the image information contents from diffraction effects.

These relations are the direct consequence of the theory of diffraction of Fraunhofer. In case the system has some optical aberrations (either static or due to the turbulence of the atmosphere), the relations still hold but the pupil function is now complex and equal to:

$$P_{ab}(u, v) = P(u, v)e^{i\phi_{ab}(u,v)} \tag{7.3}$$

where $\phi_{ab}(u, v)$ is the distribution of phase due to aberrations across the pupil.

7.2.2 Imaging with Sparse Apertures

This theory applies as well when the aperture is diluted as is the case for an interferometer. The pupil function is then the discrete sum of sub-pupils that do not overlap. The autocorrelation function is not a continuous function anymore as is the case for a single pupil but has a number of peaks: $N(N-1)$ centered on non-zero frequencies $\pm B_{ij}/\lambda$ (where B_{ij} is the distance between telescopes i and j) and 1 peak centered on 0, the sum of all individual OTFs. The high-frequency peaks are the pure interferometric parts of the OTF. The more diluted the interferometer, the larger the gaps between the peaks in the OTF. This is the main difference with single-aperture imaging: depending on the distribution of the individual pupils, the OTF may be a collection of diluted peaks with many spatial frequencies missing. The consequence is manyfold: (1) the PSF has several peaks and the direct image obtained with the interferometer is no longer directly useful to astronomical interpretation (2) the imager is no longer a low-pass filter but a collection of band-pass filters (3) the deconvolution process is difficult to apply as many spatial frequencies are missing. As a consequence, no direct useful image can be obtained with the interferometer and other techniques are required to obtain an image as described in the following sections.

The angular resolution of the interferometer (the diffraction-limited angular resolution) can be derived as in the case of a single-pupil imager: it is the reciprocal of the highest non-zero spatial frequency in the OTF. If the maximum baseline is B_{max} it is therefore λ/B_{max}. The angular resolution is approximately:

$$\frac{\theta}{1\,\text{mas}} = \frac{\lambda}{1\,\mu\text{m}} \times \frac{B_{max}}{200\,\text{m}} \qquad (7.4)$$

In the case of the VLTI where the maximum baseline is 200 m, the angular resolution in milli-arcseconds therefore scales as the wavelength in microns. It is 2 mas at 2 μm meaning most largest red supergiants are resolved.

7.3 The Nature of Interferometric Data

The exact way to measure interferometric data depends on the type of beam combination. However, one may consider that different beam combiners provide the same data for highly diluted apertures when $B \gg D$, which is usually the case for long-baseline interferometers. With this assumption, the OTF can be described by a series of peaks of negligible width and centered on the $\pm B_{ij}/\lambda$. Each baseline contributes to a positive B_{ij}/λ and to a negative one. The two peaks are actually complex-conjugated with each other and carry the same information (because the source spatial intensity distribution takes real values). By Fourier transforming Equation 7.1, one sees that the spectrum contents of the sparse interferometer image are the spatial frequency components of each B_{ij}/λ. The frequency components are the Fourier transform of the object spatial intensity distribution $\tilde{O}(u, v)$. The interferometer therefore samples the spatial spectrum of the astronomical source.

This result can also be obtained with the theory of spatial coherence. The reader is referred to Goodman (1985) for a more comprehensive discussion. In case of the two-telescope interferometer, a fringe pattern is obtained whose characteristics are given by the complex visibility. The fringe contrast is the module of the complex visibility and the fringe phase, or the distance of the white-light fringe to the Zero Optical Path Difference position, is the phase of the complex visibility (Fig. 7.1). The complex visibility is linked to the source spatial intensity distribution by the Zernike-van Cittert theorem:

$$V(u, v) = \frac{\int \int O(\alpha, \beta) e^{-2i\pi(\alpha u + \beta v)} \, du \, dv}{\int \int O(\alpha, \beta) \, du \, dv} \tag{7.5}$$

The complex visibility is the normalized source spatial spectrum and can also be expressed as:

$$V(u, v) = \frac{\tilde{O}(u, v)}{\tilde{O}(0, 0)} \tag{7.6}$$

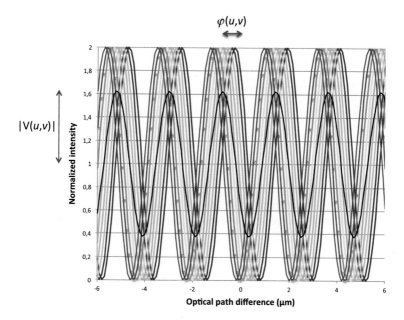

Fig. 7.1 Sum of interferograms of point-like sources in various directions. The directions are coded with colors. Interferograms on sources in different directions are shifted relative to a fixed reference. The shift is proportional to the angular offset. The resulting interferogram for an extended source is the *black* one. As the source has a non-zero angular extent, the modulus of the visibility $|V(u, v)|$ is smaller than 1. In case the source is asymmetric and resolved, the phase of the visibility $\phi(u, v)$ is no longer zero and is defined as the phase equivalent of the distance to the Zero Optical Path Difference

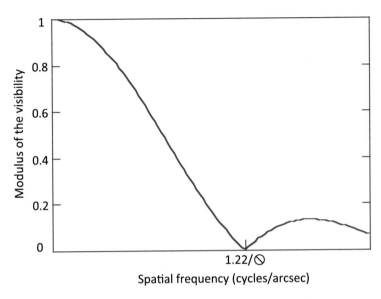

Fig. 7.2 Modulus of the visibility function of a uniform disk of diameter Ø

where the $\sim$ symbol is for the Fourier transform. The visibility is a complex number whose modulus is always smaller than 1 and whose value is exactly 1 at zero spatial frequency with the consequence that the phase of the visibility is necessarily 0 at small spatial frequencies. A classical example is the visibility function of a uniform disk of Fig. 7.2. The source spatial intensity distribution is a door function with circular symmetry $\Pi(r/\emptyset)$. Applying the Zernike-van Cittert theorem, the complex visibility is equal to:

$$V(u, v) = \frac{2J_1(\pi \emptyset S)}{\pi \emptyset S} \tag{7.7}$$

with S the modulus of the spatial frequency vector (u, v). This visibility function can be used as a ruler to measure the size of stars from a few visibility points obtained in the first lobe. As the source is centro-symmetric, the visibility function is real. It comes to a first null for exactly $S = 1.22/\emptyset$. This led to the historical method by Michelson to measure the diameter of stars by searching for the first null of the visibility function. Today, the theoretical curve is fitted to a set of points to derive the uniform disk diameter. The phase of the visibility flips from 0 to π and vice-versa after each null. The visibility of the uniform disk model is the most simple visibility function to describe the interferometric data of a star. It can be refined to account for limb darkening.

The visibility modulus is therefore to first order a proxy for the size of an object. The phase of the visibility carries very important information for imaging. The visibility is real for a centro-symmetric source, meaning the phase can only

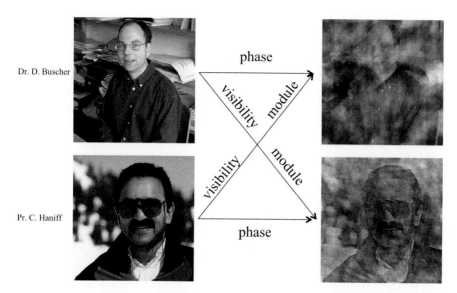

Fig. 7.3 Illustration of the importance of visibility phase for interferometric imaging (derived from Buscher (2003)). *Left*: original pictures of Dr. D. Buscher and Pr. C. Haniff. *Right*: processed images after exchanging the phase maps of the spatial spectrum or visibility. The asymmetric features are mostly coded in the phase. Images are courtesy of Dr. D. Buscher and Pr. C. Haniff

be: $\phi(u, v) = 0 \bmod(\pi)$. But it can take any value between 0 and π for a source with no particular symmetry. The visibility phase is therefore a strong indicator of asymmetries in the source spatial intensity distribution. Figure 7.3 is a classical example derived from Buscher (2003) to show the importance of visibility phase to reconstruct images in interferometry. The pictures of Doctor David Buscher and Professor Christopher Haniff from the university of Cambridge are Fourier transformed into visibilities and visibility phases are swapped. Images are reconstructed (inverse Fourier transform) using the visibility modulus and the swapped visibility phase maps. One can easily distinguish the picture of each colleague in these synthetic images showing that visibility phases carry important informations for imaging.

Since the visibility function is equal to 1 at 0 spatial frequency, the phase can only be 0 for short telescope spacings. This means that all sources tend to look the same and look symmetric when they are not resolved. The source has to be resolved to be able to measure useful phase information. In practice phases can take non-zero values below 50 % of visibility modulus. This fact can be used as a calibration tool for phases.

7.4   Aperture Synthesis and Supersynthesis

One of the most striking characteristics of optical/infrared interferometers is the relatively small number of telescopes. There are several reasons to this, one of which is the increasing complexity of the interferometer with the number of telescopes as the beams need to be carried down to a beam combination station without loss of spatial coherence and then need to be recombined into $N(N-1)$ baselines. An additional reason in the case of large telescopes is their cost. The major consequence of this fact is the sparseness of the array: the (u, v) plane is sampled with very few peaks.

The original idea behind the aperture synthesis technique is that a large pupil can be built with many small ones. This is a very naive idea indeed as one single large pupil filled with small ones would lead to redundant baselines: different pairs of telescopes would yield the same baselines and therefore the same spatial information as they would sample the same spatial frequencies. This is illustrated in Fig. 7.4 where 6 telescopes are recombined in two different ways. First, by trying to fill a single large pupil, in this case the number of independent spatial components is equal to 9 (plus the central peak that corresponds to a single-telescope OTF). Second, by trying to maximize the number of sampled spatial frequency components while keeping the interferometer OTF as compact as possible. This is the Golay 6 distribution of pupils whose effect is to provide exactly 15 independent spatial frequencies while keeping the interferometer OTF compact. This example shows that synthesizing a pupil is not what matters but what is important is to synthesize the best OTF possible instead by maximizing the number of spatial frequencies provided by the interferometer.

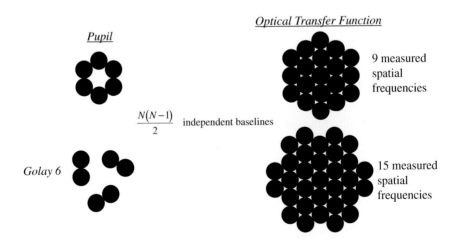

Fig. 7.4 Redundant and non-redundant configurations. *Left*: two different pupil setups with 6 apertures. *Right*: the supports of the *OTF*. The Golay 6 configuration leads to 6 high-frequency additional peaks in the *OTF*

In the above example, compactness was a major constraint. It leads to a limited angular resolution. It is not necessarily absolutely mandatory to measure all spatial frequencies up to a cut-off frequency. If this constraint is released then the angular resolution can be higher. But, the (u, v) plane sampling has to be rich enough to minimize the number of unknowns for image reconstruction (see Sect. 7.6). In sparse interferometers, the *OTF* or (u, v) plane coverage is also sparse. Interferometers use the rotation of Earth to increase the number of sampled visibilities. This technique is called *supersynthesis* and builds upon the fact that for each baseline of the interferometer the spatial frequency at which visibilities are measured is the projection of the baseline in the direction of the source divided by the wavelength. As a consequence, for a given telescope pair, the spatial frequency will vary with time along a (u, v) track whose characteristics depend upon the source declination, the hour angle and the latitude of the interferometer. The equation of (u, v) tracks is given by (see e.g. Ségransan 2003):

$$u^2 + \left(\frac{v - (Bz/\lambda) \cos \delta}{\sin \delta} \right)^2 = \frac{B_x^2 + B_y^2}{\lambda} \tag{7.8}$$

(u, v) tracks are ellipses whose center is on the v axis. (u is pointing towards East, and v towards North). B_x, B_y and B_z are the coordinates of the baseline vector projected onto the axes pointing towards East, North and the meridian, respectively. Two particular cases can be underlined:

- $\delta = 0^o$: (u, v) tracks are straight lines parallel to the u axis;
- $\delta = 90^o$: (u, v) tracks are circles centered on the origin.

An example of (u, v) tracks at VLTI is presented in Fig. 7.5. This example is a good mix between supersynthesis and the use of various telescopes or telescope

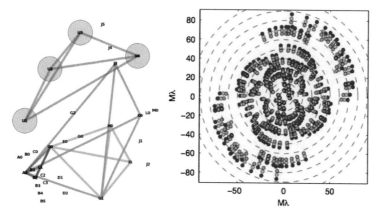

Fig. 7.5 (u, v) tracks (*left*) obtained at VLTI with (*right*) 1 configuration with the 4 Unit Telescopes, 2 configurations with 4 Auxiliary Telescopes and 1 configuration with 3 Auxiliary Telescopes. Courtesy of Dr. Jean-Philippe Berger

positions at VLTI: the rich (u, v) plane was obtained by using 3 quadruplet configurations (2 with the Auxiliary Telescopes and 1 with the Unit Telescopes) and 1 triplet configuration (with the Auxiliary Telescopes), each providing 6 or 3 baselines, and by improving the (u, v) plane coverage through Earth rotation. This (u, v) plane coverage is quite dense and circularly symmetric (except at the highest spatial frequencies though) and is one of the best examples of spatial frequency contents obtained in optical/infrared interferometry so far.

7.5 Closure Phases

Optical/infrared interferometers are very sensitive to atmospheric turbulence. The main effect is the degradation of spatial coherence: the module of the visibility scales with the coherent energy (Rousset et al. 1991) available in the pupil $e^{-\sigma_\phi^2}$. The variance of phase σ_ϕ^2 increases rapidly with the diameter of the pupils: $\sigma_\phi^2 = 1.03 \left(\frac{D}{r_0} \right)^{\frac{5}{3}}$, where r_0 is the Fried parameter (the seeing is $\frac{\lambda}{r_0}$), causing the fringe contrast to drop with telescope diameter D. The coherent energy is improved in interferometers using pupils larger than r_0 by correcting turbulence with adaptive optics. The combination of adaptive optics or small pupils with modal filtering with fibers or integrated optics allows to restore very high fringe contrasts with a high accuracy (see examples in Perrin et al. 1998 or Lacour et al. 2008).

Atmospheric turbulence and other phase errors have an another important impact on the measurement of visibility phases. The phase of the visibility is the normalized distance measured between the zero optical path difference (ZOPD) position and the white light or central fringe of the interferogram (Sect. 7.3). The ZOPD position is disturbed by the piston effect whose cause is illustrated in Fig. 7.6. The piston is the spatial average phase in the pupil of the telescope. Different telescopes have different pistons (the correlation decreases with the distance between telescopes) and interferometers are sensitive to the differential piston between pupils. The differential piston shifts the phase of the interferogram by $\Delta\phi$ or induces a motion of the ZOPD of $\frac{\lambda}{2\pi}\Delta\phi$. At optical/infrared wavelengths, the shift is many fringes as shown in Fig. 7.7 where the fringe shift was measured with the PRIMA fringe tracker of the VLTI as a function of time (Sahlmann et al. 2009). The piston effect correlation time is small (a few tens of milli-seconds depending on seeing conditions) and forbids any long integration time if not compensated in real time with a fringe tracker. The non-stationarity of turbulence and the large amplitude of piston also prevent in practice to average it down to zero. The ZOPD exact position is therefore lost and it is not possible to measure absolute values of the phase of visibilities. This is a very strong consequence as half of the interferometric information (the one on asymmetries) is lost.

This was a historical problem for radio interferometry also although to a lesser extent. The closure phase technique was invented to solve this issue (Jennison 1958). The principle is very simple. Each telescope contributes an error ϵ_i on phase so that

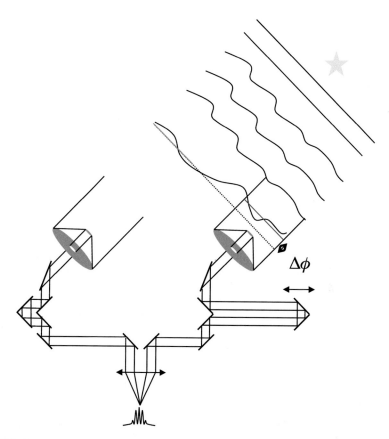

Fig. 7.6 Sketch of a 2-telescope interferometer showing the nature of differential piston. The piston mode is the spatial average of phase in a telescope pupil. Differential piston is the difference of average phase between 2 pupils. Differential piston causes fringe motion at the beam combination point of the interferometer and can be compensated with fast delay lines

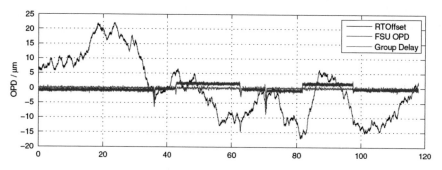

Fig. 7.7 Measurement of fringe shift due to differential piston with the PRIMA fringe tracker of the VLTI as a function of time (Sahlmann et al. 2009)

the visibility phase measured for baseline ij writes:

$$\phi_{ij}^{obs} = \phi_{ij} + \epsilon_i - \epsilon_j \qquad (7.9)$$

The errors can be smartly eliminated by summing the observed visibility phases over a triangle of baselines:

$$\phi_{12}^{obs} = \phi_{12} + \epsilon_1 - \epsilon_2$$
$$\phi_{23}^{obs} = \phi_{23} + \epsilon_2 - \epsilon_3$$
$$\phi_{31}^{obs} = \phi_{31} + \epsilon_3 - \epsilon_1$$

- - - - - - - - - - - - - - -

$$\sum \phi_{ij}^{obs} = \sum \phi_{ij} \qquad (7.10)$$

Closure phases are therefore immune to piston errors or other phase perturbations and are pure observables as they only depend on the geometry of the object. Closure phases were first demonstrated on long baselines by the Cambridge group with the COAST interferometer (Baldwin et al. 1996).

One can demonstrate that closure phases convey similar informations on the asymmetries of the source as visibility phases. The closure phases are equal to $0 \mod(\pi)$ and can take any value between 0 and π otherwise. The amount of asymmetric flux in the spatial intensity distribution of a source is directly linked to the closure phase value if the source is resolved by the interferometer (Monnier et al. 2007):

$$\text{Closure phase (rad)} \simeq \frac{\text{asymmetric flux}}{\text{symmetric flux}} \qquad (7.11)$$

Based on Huby et al. (2012) and following the theoretical derivation by Baldwin and Haniff (2002), the $1\,\sigma$ dynamic range in a reconstructed image scales with the accuracy on closure phases:

$$DR \simeq 88\sqrt{\#B}\left(\frac{1^o}{\sigma_{CP}}\right) \qquad (7.12)$$

with $\#B$ the number of independent baselines or spatial frequencies and σ_{CP} the accuracy on closure phases. This theoretical formula is only indicative. The dynamic range in a reconstructed image depends on other factors such as the noise on visibility moduli and on the exact (u, v) coverage but this gives a hint of what could be expected. With 4 different configurations of a 4-telescope interferometer a 1° accuracy on closure phases translates into a 1:400 dynamic range to detect spots at the surface of a star. The best accuracy demonstrated on closure phases reaches just a few $\sim 0.1^o$ (see e.g. Lacour et al. 2011) meaning that spots could be imaged if they contain a few thousandths of the star flux.

Closure phases are not the ideal solution to the issue of piston though. As a matter of fact, for a 3-telescope interferometer, 3 visibility phases have to be measured to recover the full phase information. A single closure phase is measured instead meaning that only 33 % of the phase information can be obtained. But this is to be compared to no phase information at all and the gain is infinite. More generally, the amount of phase information obtained with the closure phase technique can be derived by counting the number of independent triangles with a set of N telescopes which is equal to $\frac{(N-1)(N-2)}{2}$. This has to be compared to the number of baselines $\frac{N(N-1)}{2}$. The fraction of phase information is therefore: $1 - \frac{2}{N}$. It increases with the number of telescopes and reaches 80 % with 10 telescopes or $\simeq$ 90 % with 21 telescopes as is the case for the VLA. With a 4-telescope interferometer like VLTI, this fraction is equal to 50 %, 67 % with the CHARA array with 6 telescopes. This would be insufficient if all visibility phases were a priori unknown. However, depending on the distribution of baselines relative to the source geometry, a more or less large fraction of the visibility phases can be set to an a priori value. As recalled in Sect. 7.3, the phase of the visibility function is necessarily zero for short baselines. Partial redundancy in the array can also reduce the number of unknowns. All in all, the fraction of unknown phases is therefore smaller than given than the formula above. The full information has to be recovered though to reconstruct an image. The principle is described in the next section.

7.6 Image Reconstruction

Image reconstruction is a difficult problem in interferometry but is not unique and has some similarities with other problems in other fields. In Sect. 7.2.2 we discussed an interferometer as a collection of low-pass filters. The visibilities are filtered by the *OTF* or, equivalently, the spatial intensity distribution is convolved with a multiple-peak *PSF* called the dirty beam by radio astronomers. The problem is therefore equivalent to a deconvolution problem as encountered in imaging with adaptive optics for example but with a complex *PSF*. Another way to see it is to consider an image as a model (Fig. 7.8). The image is a set of $N \times N$ pixels with as many fluxes to be determined. The parameters are therefore the N^2 pixel fluxes. The problem is all the more difficult as the number of parameters/pixels is large. The sparseness of optical arrays adds even more to these difficulties as the number of collected informations is in practice limited and smaller than the number of pixels to reconstruct. Image reconstruction in optical/infrared interferometry is a really complex problem.

Several methods have been developed or are used to reconstruct images in (optical/infrared) interferometry of which one may cite a few examples:

- CLEAN;
- Maximum likelihood;
- χ^2 with regularization;

Fig. 7.8 Sketch of a 16×16
pixel image. From image
reconstruction point of view,
the image is like a
256-parameter model, the
pixel intensities, whose value
need to be constrained from
the interferometric data and
from a priori informations
(regularization)

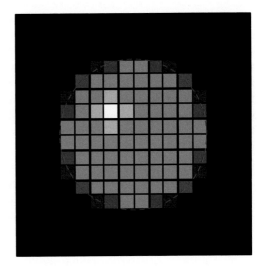

- Maximum entropy method;
- Building block method.

I will only give here the principles of the χ^2 method with regularization which
is discussed in more details in Thiébaut (2013). It is derived from the classical χ^2
method to constraint the parameters of a model applied to interferometric data:

$$\chi^2(I_m) = \frac{1}{M-N^2} \left[\sum_{S_i} \left(\frac{|V_i|^2 - |V_{I_m}|^2(S_i)}{\sigma_{|V_i|^2}} \right)^2 \right] \tag{7.13}$$

$$+ \frac{1}{M-N^2} \left[\sum_{S_i,S_j} \left(\frac{CP_{i,j} - CP_{I_m}(S_i,S_j))}{\sigma_{CP_{i,j}}} \right)^2 \right]$$

where M is the total number of data ($|V|^2$ and closure phases, squared quantities
are used for the modulus of the visibility instead of the linear quantity because
of the existence of an estimator immune to noise bias), N^2 is the size of the
reconstructed image in pixels, $|V_i|^2$ and $CP_{i,j}$ are the squared modulus visibilities
and the closure phases data, $|V_{I_m}|^2(S_i)$ and $CP_{I_m}(S_i, S_j)$ are the equivalent for the
model/image at spatial frequencies S_i and (S_i,S_j) for the closure phases (closure
phases are defined for a triangle of spatial frequencies which is fully defined
with two spatial frequencies), $\sigma_{|V_i|^2}$ and $\sigma_{CP_{i,j}}$ are the respective estimated errors.
A simpler version of image reconstruction is called *parametric imaging*. The χ^2
function can be minimized against an image model, i.e. a 2D astrophysical model,
with a few parameters (for example the diameter of the star, the limb darkening
coefficient(s) and some spot parameters). The image in this case is pre-determined

by the model. Examples of *parametric imaging* are given in Lacour et al. (2008) or Haubois et al. (2009).

For *blind image reconstruction*, i.e. without an a priori model of the image, the formula has to be refined to take extra constraints into account. As a matter of fact, in the image reconstruction process, some information has to be produced at spatial frequencies where no data were taken. The solution cannot be unique and many possibilities satisfy the minimum χ^2 criterion. The uniqueness of the solution can be forced by adding an extra penalty term to the χ^2 which forces the solution to be compatible with some a priori information on the object and called the prior. Equation 7.13 then becomes:

$$\chi^2(I_m) = \frac{1}{M-N^2}\left[\sum_{S_i}\left(\frac{|V_i|^2 - |V_{I_m}|^2(S_i)}{\sigma^2_{|V_i|^2}}\right)^2\right] \tag{7.14}$$

$$+\frac{1}{M-N^2}\left[\sum_{S_i,S_j}\left(\frac{CP_{i,j} - CP_{I_m}(S_i,S_j))}{\sigma^2_{CP_{i,j}}}\right)^2\right]$$

$$+\mu \times \text{penalty function}$$

The penalty function is a regularization term which adds constraints to reconstruct the image (e.g. positivity, smoothness, limited extension, general shape ...). The hyper-parameter μ can be adjusted to weigh the relative influences of the data and of the a priori information. The higher μ, the closer the final image to the prior. The reconstructed image is therefore not only determined by the data but also by the choice of the hyper parameter and by the choice of the constraints. Let us assume that the penalty function is a simple squared distance to a particular image, if μ tends towards infinity then the final reconstructed image will be the same as the particular image, whatever the data. This method is very powerful to reconstruct images with physical significance but one has to be cautious with the tuning parameters and aware that the particular solution provided by the image reconstructor is determined by extra informations independent of the data. Image reconstruction in optical/infrared interferometry is therefore difficult and primarily because of the lack or sparseness of the observed information. The characteristics of the reconstructed image can be compared with some characteristics directly derived from the visibilities to check the relevance of the image and make sure the main features of the image are compatible with the most certain information included in the data. Example of reconstructed images based on long-baseline optical/interferometer data are given in the next section.

7.7 Examples of Results

Since the first image of the Capella binary system obtained by Baldwin et al. (1996), most interferometric studies have used closure phases to detect asymmetries without reconstructing images because of poor (u, v) coverage. The situation has changed in the second half of the 2000 decade during which both interferometers and image reconstruction techniques applied to the optical/infrared domain have improved.

Parametric imaging is a powerful tool when the geometry of the object is likely to be simple. A very spectacular example is the image of Vega obtained with NPOI at 500 nm (Peterson et al. 2006). This image is a wonderful evidence that Vega is a rapid rotator (gravitational darkening) seen almost pole on (Fig. 7.9), thus providing a wonderful explanation why Vega could not be identified as a particular star by classical techniques without spatial information. Another one is the evolution of the photosphere of Betelgeuse in the H band observed by VLTI/PIONIER between 2012 and 2014 (Montargès et al. 2015). The data are quite sparse for image reconstruction and the (u, v) plane coverage mostly samples a single direction. However a feature is clearly detected in the closure phase data and can be fitted as a single spot (Fig. 7.10). These are complemented with new data in 2015 whose processing is on going as well as the overall interpretation of the full sequence of images.

Betelgeuse was observed in the H band too by Haubois et al. (2009) with the IONIC instrument at IOTA in 2005. The (u, v) plan coverage was more dense

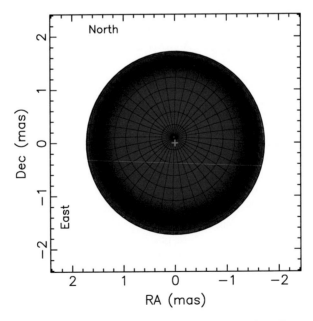

Fig. 7.9 Parametric image of Vega obtained with the NPOI array at 500 nm (Peterson et al. 2006). This image is a wonderful evidence that Vega is a rapid rotator (gravitational darkening) seen almost pole on

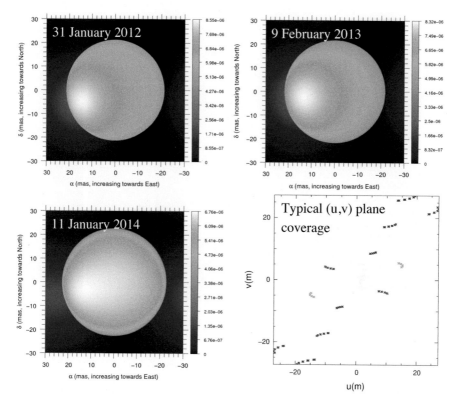

Fig. 7.10 Parametric images of Betelgeuse obtained with the VLTI/PIONIER instrument in the H band with (u, v) plane coverage (Montargès et al. 2015)

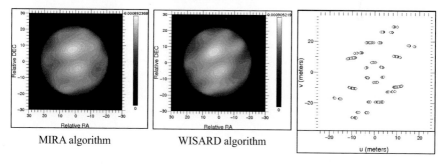

Fig. 7.11 Images of Betelgeuse obtained with the IOTA/IONIC instrument in the H band with (u, v) plane coverage (Haubois et al. 2009). The two images were obtained with two different algorithms, MIRA and WISARD

in this case and could allow to reconstruct an image in parallel with parametric imaging. The image was reconstructed with two different codes: MIRA (Thiébaut 2008) and WISARD (Meimon et al. 2009). As Fig. 7.11 shows, the two different algorithms lead convincingly to the same result to within noise (noise from the data

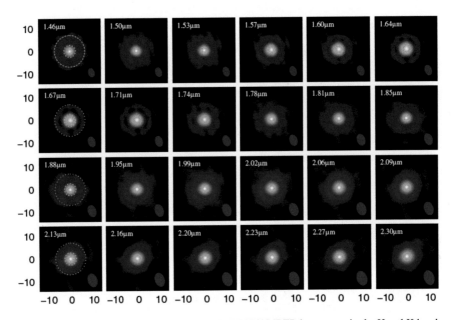

Fig. 7.12 Images of T Leporis obtained with the VLTI/AMBER instrument in the H and K bands (Le Bouquin et al. 2009). The *dashed circles* are respectively the sizes of the star and of the CO and H_2O molecular shell in the astrophysical model. The axes are in mas units

and image reconstruction noise). Both algorithms are based on the regularization method applied to the χ^2 analysis. The best limb-darkened fit of the star with an environment was used as the prior. In both cases a quadratic regularization was used meaning that the strong intensity gradients between the a priori image and the reconstructed image are quadratically minimized.

The main information in the image is the presence of two spots, one barely resolved and one unresolved. It is hard to tell if other features are real but a large fraction of them is probably of noisy origin. This was the first evidence of the presence of spots at the surface of Betelgeuse in this wavelength range and was demonstrated to be consistent with the presence of convective cells (Chiavassa et al. 2010). This triggered the first successful search for a magnetic field in a red supergiant attributed to a dynamo effect in the convective cells (Aurière et al. 2010).

Le Bouquin et al. (2009) have produced the first image of a star observed with the VLTI with the AMBER instrument in the H and K bands (Fig. 7.12). It is a series of images obtained in various spectral channels from 1.46 to 2.30 μm. The MIRA software was used for image reconstruction. Two steps were applied with two different regularizations: smoothness and quadratic. T Leporis is a Mira star and the images clearly show a mostly centro-symmetric object surrounded by a shell with CO and H_2O absorption, consistent with previous observations of the same type of objects. Mira stars can therefore be centro-symmetric to a high degree!

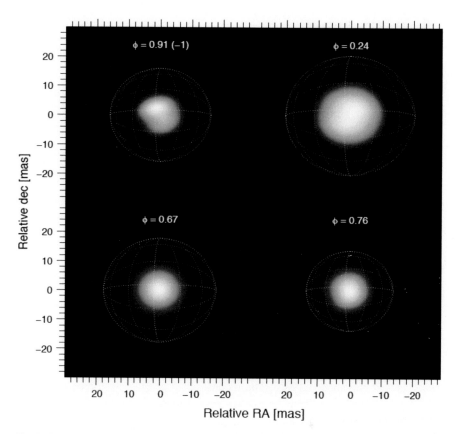

Fig. 7.13 Images of χ Cygni obtained with the IOTA/IONIC instrument in the H band (Lacour et al. 2009). The *dashed sphere* is the molecular shell derived from the modeling of the data

Lacour et al. (2009) have observed another Mira star, χ Cygni, at four different epochs in the H band with IONIC at the IOTA interferometer. The images were reconstructed with MIRA using a limb-darkened disk as the prior and a weighted quadratic distance to the prior as regularization term (see Fig. 7.13). Not all images are centro-symmetric despite the prior and asymmetric features are clearly detected. The star is all the more asymmetric as its diameter is small during the pulsation cycle. This could be a signature of convection, this has to be confirmed by hydrodynamic simulations. Contrary to the T Leporis images, no molecular shell is visible around the central star although it is necessary to fit the visibilities as demonstrated in the paper by parametric images. The shell in the case of χ Cygni seems more detached and thinner in angular width, a difficulty for the regularization algorithm. The images in this case were of great use to find the best model for parametric imaging which provided the basis for the astrophysical analysis.

The last example of image reconstruction from near-infrared data, and to close the loop with Vega, is the work by Monnier et al. (2007) on Altair. Altair is a rapid

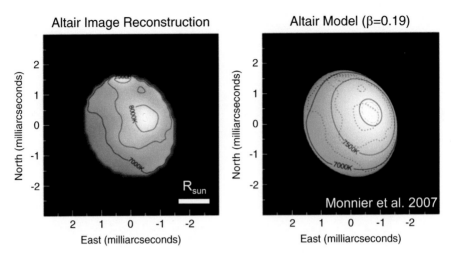

Fig. 7.14 *Left*: Image of Altair obtained with the CHARA/MIRC instrument (4 telescopes) in the H band (Monnier et al. 2007). *Right*: gravity darkening model of the source superimposed on the reconstructed image isocontours

rotator as well. The image was obtained with the MIRC instrument at CHARA using 4 telescopes (Fig. 7.14). They used a different algorithm than the regularized χ^2: the MACIM algorithm (Ireland et al. 2006) based on the Maximum Entropy Method (Narayan and Nityananda 1986). The reconstructed image was forced to fit into an a priori ellipse, the ellipse parameters were obtained from the interferometric data. The goal of the work was to study the surface of the star and in particular gravity darkening which is expected in the case of rapid rotators. The images clearly show that the pole is brighter than the equator. The latitude dependence of gravity darkening is described by the β parameter. A value of 0.19 is derived from the data whereas the von Zeipel theory prediction is 0.25. These data were a fundamental contribution to the discussion and improvement of the von Zeipel theory with new fast rotator theories and models.

7.8   Conclusion

After the pioneering image of Capella by the COAST interferometer in 1996, imaging with interferometers is now becoming more commonplace. A few arrays working at optical and infrared wavelengths produce images of stellar surfaces. Asymmetries or spots are detected for a large fraction of them. This will certainly contribute to a new era of stellar physics.

Acknowledgements I am grateful to the organizers of the Besançon school for their invitation.

References

Aurière, M., Donati, J. -F., Konstantinova-Antova, R., Perrin, G., Petit, P., & Roudier, T. (2010). *Astronomy and Astrophysics, 516,* L2.

Baldwin, J. E., Beckett, M. G., Boysen, R. C., Burns, D., Buscher, D. F., Cox, G. C., et al. (1996). *Astronomy and Astrophysics, 306,* L13.

Baldwin, J. E., & Haniff, C. A. (2002). *Philosophical Transactions of the Royal Society of London, Series A, 360*(1794), 969.

Buscher, D. F. (2003). *Proceedings of the School Observing the Very Large Telescope Interferometer.* In G. Perrin & F. Malbet (Eds.), *EAS Publications Series: Vol. 6* (p. 227).

Chiavassa, A., Haubois, X., Young, J. S., Plez, B., Josselin, E., Perrin, G., et al. (2010). *Astronomy and Astrophysics, 515,* A12.

Danchi, W. C., Bester, M., Degiacomi, C. G., Greenhill, L. J., & Townes, C. H. (1994) *Astronomical Journal, 107,* 1469.

Goodman, J. W. (1985). *Statistical optics.* New York: Wiley.

Haubois, X., Perrin, G., Lacour, S., Verhoelst, T., Meimon, S., Mugnier, L., et al. (2009). *Astronomy and Astrophysics, 508,* 923.

Huby, E., Perrin, G., Marchis, F., Lacour, S., Kotani, T., Duchêne, G., et al. (2012). *Astronomy and Astrophysics, 541,* A55.

Ireland, M., Monnier, J. J. D., & Thureau, N. (2006). *Society of Photo-Optical Instrumentation Engineers (SPIE) Conference Series: Vol. 6268*

Jennison, R.C. (1958). *Monthly Notices of the Royal Astronomical Society, 118,* 276.

Josselin, É, & Plez, B. (1997). *Astronomy and Astrophysics, 469,* 671.

Kervella, P., Verhoelst, T., Ridgway, S. T., Perrin, G., Lacour, S., Cami, J., et al. (2009). *Astronomy and Astrophysics, 504,* 115.

Lacour, S., Thiébaut, E., Perrin, G., Meimon, S., Haubois, X., Pedretti, E., et al. (2009). *Astrophysical Journal, 707,* 632.

Lacour, S., Meimon, S., Thiébaut, E., Perrin, G., Verhoelst, T., Pedretti, E., et al. (2008). *Astronomy and Astrophysics, 485,* 561.

Lacour, S., Tuthill, P., Amico, P., Ireland, M., Ehrenreich, D., Huelamo, N., et al. (2011). *Astronomy and Astrophysics, 532,* A72.

Le Bouquin, J. -B., Lacour, S., Renard, S., Thiébaut, E., Merand, A., & Verhoelst, T. (2009). *Astronomy and Astrophysics, 496,* L1.

Meimon, S., Mugnier, L. M., & Le Besnerais, G. (2009). *Journal of the Optical Society of America A, 26,* 108.

Monnier, J. D., Zhao, M., Pedretti, E., Thureau, N., Ireland, M., Muirhead, P., et al. (2007). *Science, 317,* 342.

Monnier, J. D. (2007). *New Astronomy Reviews, 51,* 604.

Montargès, M., Kervella, P., Perrin, G., Chiavassa, A., Le Bouquin, J. -B., Ridgway, S. T., et al. (2016). The close circumstellar environment of Betelgeuse III. VLTI/PIONIER interferometric monitoring of the photosphere. *Astronomy & Astrophysics* (submitted).

Narayan, R., & Nityananda, R. (1986). *Annual Review of Astronomy and Astrophysics, 24,* 127.

Ohnaka, K., Hofmann, K. -H., Benisty, M., Chelli, A., Driebe, T. Millour, F., et al. (2009). *Astronomy and Astrophysics, 503,* 183.

Perrin, G., Coudé du Foresto, V., Ridgway, S. T., Mariotti, J.-M., Traub, W. A., Carleton, N. P., et al. (1998). *Astronomy and Astrophysics, 331,* 619.

Peterson, D. M., Hummel, C. A., Pauls, T. A., Armstrong, J. T., Benson, J. A., Gilbreath, G. C., et al. (2006). *Nature, 440,* 896.

Richichi, A., Percheron, I., & Khristoforova, M. (2005). *Astronomy and Astrophysics, 431,* 773.

Rousset, G., Madec, P.-Y., & Rabaud, D. (1991). *High-Resolution Imaging by Interferometry II, No. 39 in ESO Conference and Workshop Proceedings* (p. 1095).

Sahlmann, J., Ménardi, S., Abuter, R., Accardo, M., Mottini, S., & Delplancke, F. (2009). *Astronomy and Astrophysics, 507,* 1739.

Ségransan, D. (2003). *Proceedings Of The School Observing The Very Large Telescope Interferometer*. In G. Perrin, & F. Malbet (Eds.), *EAS Publications Series: Vol. 6* (p. 69).

Thiébaut, É. (2008). *Proceeding of the Society of Photo-Optical Instrumentation Engineers (SPIE) Conference Series: Vol. 7013* (p. 43).

Thiébaut, É. (2013). *EAS Publications Series, 59*, 157.

Chapter 8
Interferometric Surface Mapping of Rapidly Rotating Stars: Application to the Be star Achernar

Armando Domiciano de Souza

Abstract Rotation is one of the fundamental parameters that governs the physical structure and evolution of stars. Massive stars are those presenting the highest rotation velocities and thus those for which the consequences of rotation are the strongest. On the stellar photosphere fast-rotation induces (1) a geometrical flattening and (2) a non-uniform distribution of flux/effective temperature (gravity darkening effect). A detailed mapping of these effects on the stellar photosphere, including large scale surface velocity fields, is nowadays possible thanks to modern techniques of optical/infrared long-baseline interferometry (OLBI). In this paper we focus on the measurement of gravity darkening from OLBI, while the determination of flattening is detailed by Kervella (this volume). In addition, we also show that, for fast-rotators, the combination of OLBI and spectroscopy (spectro-interferometry) allows to go beyond the spatial resolution limit of interferometers in order to measure angular sizes of stars, otherwise not measurable by classical OLBI techniques. The results presented here are based on ESO-VLTI interferometric observations of the Be star Achernar.

8.1 Introduction

Rotation is present everywhere in astrophysics, from the smallest bodies of the solar system to the largest scale structures in the Universe. In stars, rotation is present since their formation from the interstellar molecular clouds, during all the protoplanetary phase, main sequence, formation of planetary systems, and up to the final stages of stellar evolution, playing a crucial role for example in supernovae explosions, formation of pulsars and rotating black holes, possible origin of some gamma-ray bursts.

A. Domiciano de Souza (✉)
Laboratoire Lagrange, Université Côte d'Azur, Observatoire de la Côte d'Azur, CNRS,
Boulevard de l'Observatoire, CS 34229, 06304 Nice cedex 4, France
e-mail: Armando.Domiciano@oca.eu

© Springer International Publishing Switzerland 2016
J.-P. Rozelot, C. Neiner (eds.), *Cartography of the Sun and the Stars*,
Lecture Notes in Physics 914, DOI 10.1007/978-3-319-24151-7_8

The first measurements proving that stars rotate started 400 years ago with the Galilei Galileo observations of mysterious dark features moving in front of the visual image of the Sun. After some debates, these features were correctly interpreted as dark spots on the solar surface and their movement was caused by the rotation of the Sun. Since then, this field encountered a initially slow development, probably due to the lack of realistic physical models and precise observational data. Nowadays, in particular thanks to many theoretical and observational results obtained over the twentieth century, rotation is considered a fundamental stellar parameter that, together with the stellar mass and metallicity, governs the structure and evolution of stars across the Hertzsprung–Russell (H-R) diagram.

Of course, the effects of rotation on stars are stronger for stars with high rotation velocities. Modern stellar interferometers are now currently used to map the photosphere of fast-rotating stars. An overview of optical/infrared long-baseline interferometry (OLBI) results on rapid rotators from 2001 to 2011 is given by van Belle (2012). The results obtained include stellar parameters derived from physical models as well as interferometric reconstructed images of the stellar photosphere.

Among the fast-rotators, the Be stars[1] are those presenting the highest rotation velocities among non-degenerate stars. Achernar (α Eridani, HD 10144; spectral type B6Vep) is the closest ($d = 42.75 \pm 1.04$ pc; van Leeuwen 2007) and brightest Be star as seen from Earth. It is easily observable in the southern hemisphere, being the ninth star in the night sky in visible light (V = 0.46). As discussed by Kervella (these proceedings), Achernar has a strong rotational flattening due to its fast rotation (Domiciano de Souza et al. 2003; Kervella and Domiciano de Souza 2006). As we show in the following, these characteristics make Achernar an ideal target to study fast-rotation effects on the stellar surface using OLBI techniques.

8.2 Quick Overview of OLBI

In this section we give some basic concepts of OLBI techniques, which were used to obtain the results presented in this work. The capability of OLBI to map the surface of stars is based on a relation between the observed interferometric fringes (translated into complex visibilities) and the sky-projected monochromatic brightness distribution (also called intensity map of the apparent stellar surface). Thus, for a given intensity map I_λ, the Van Cittert-Zernike theorem (e.g. Born and Wolf 1999) relates the observed complex visibility to the Fourier transform (FT) of I_λ, normalized by its value at the origin, i.e.,

$$V(u, v, \lambda) = |V(u, v, \lambda)| \, e^{i\phi(u,v,\lambda)} = \frac{\tilde{I}_\lambda(u, v)}{\tilde{I}_\lambda(0, 0)} , \qquad (8.1)$$

---

[1] A non-supergiant B star whose spectrum has, or had at some time, one or more Balmer lines in emission.

where $\tilde{I}_\lambda$ is the FT of I_λ, given explicitly by

$$\tilde{I}_\lambda(u, v) = \iint I_\lambda(x, y) e^{-i2\pi(xu+yv)} dx dy . \tag{8.2}$$

Note that the normalization term $\tilde{I}_\lambda(0, 0)$ is the stellar flux. The amplitude $|V(u, v, \lambda)|$ and phase $\phi(u, v, \lambda)$ of the complex visibility are related, respectively, to the contrast and position of the observed interferometric fringes. The spatial (angular) coordinates x and y describe the position of the intensity map on the sky-plane and are generally chosen to follow the coordinates of right ascension (α) and declination (δ). They are defined by the actual (linear) position on the visible stellar surface projected onto the sky-plane, divided by the distance d to the star. Of course, OLBI does not provide observations at all spatial (or Fourier) frequencies, u and v, but only at points corresponding to the actual separations (baselines) of telescopes, i.e.,

$$(u, v) = \frac{B_{\text{proj}}}{\lambda} = \frac{B_{\text{proj}}}{\lambda}(\sin \text{PA}, \cos \text{PA}) , \tag{8.3}$$

where B_{proj} is the baseline (distance between two telescopes) projected onto the target's direction, and PA is the position angle of the baseline (from north to east). Observations with different baselines and/or at different times during the night (technique of Earth-rotation synthesis) are often used to increase the number of observed points in the uv-plane, i.e., the uv-coverage.

The instrument (telescopes, optics, detectors, etc.) and the Earth atmosphere alter the recorded target light in different ways that cannot be completely compensated. As a consequence, the observed complex visibility is given by

$$V_{\text{obs}}(u, v, \lambda, t) = V(u, v, \lambda) OTF(u, v, \lambda, t)$$
$$= V(u, v, \lambda) |OTF(u, v, \lambda, t)| e^{i\Delta\varphi(u,v,\lambda,t)} , \tag{8.4}$$

where $V(u, v, \lambda)$ is given defined in Eq. (8.1) and OTF is the complex optical transfer function, which is the FT of the combined time-dependent point spread function (PSF) of the instrument and of the Earth atmosphere. Note that the target can also vary in time, but we are interested here only in the temporal variations of the OTF and its effects on the interferometric observables.

In the following we describe some techniques allowing to (at least partially) estimate the actual complex visibility V of the target (amplitude and phase) from the observed one V_{obs}.

8.2.1　Visibility Amplitudes

To estimate the visibility amplitude $|V(u, v, \lambda)|$ from observed fringe contrasts ($|V_{obs}|$) it is thus often necessary to combine the observations of the target with those of a calibrator star of known size (and thus of known theoretical fringe contrast). The observations of target and calibrator need to be performed close in time and with the same instrumental conditions. This procedure allows to partially compensate some effects that corrupt the measurements of interference fringes, diminishing the fringe contrast.

Indeed, one can see from Eq. (8.4) that the observed visibility amplitude $|V_{obs}|$ is given by

$$|V_{obs}(u, v, \lambda)| = |V(u, v, \lambda)| |\langle OTF \rangle_t| = |V(u, v, \lambda)| T , \qquad (8.5)$$

where the proportionality factor T is called (modulus) transfer function (also represented as MTF), which is the modulus of the OTF averaged over the time span of the observations. The observation of calibrators of known sizes thus allows to estimate T from their observed $|V_{obs}^{cal}|$ and theoretical $|V^{cal}|$ visibility amplitudes. The estimated T can then be applied to the target, preferably interpolating between observations of calibrators performed before and after those of the target.

8.2.2　Differential and Closure Phases

Because of instrumental instabilities and optics imperfections, and because of the fast and unpredictable phase variations introduced in the light path by the turbulence of Earth atmosphere, interferometers cannot measure directly the target's FT phase $\phi(u, v, \lambda)$ (Eq. (8.1)). However, at least part of the phase information can be recovered from two observable quantities related to $\phi(u, v, \lambda)$, namely, the *differential phase* and the *closure phase*.

The *differential phase* is essentially $\phi(u, v, \lambda)$ as a function of wavelength from which a reference phase calculated at a chosen wavelength (or wavelength range) λ_{ref} has been subtracted. Somewhat different ways of estimating the differential phase from observations exist, depending on the instrument used and on the assumptions made for the calculations. Since the results using differential phases in this work are based on the observations from the VLTI/AMBER beam combiner (Petrov et al. 2007), the formalism adopted here for the differential phase estimations is similar to the one described by Millour et al. (2011, 2006) for this instrument.

The time-average of the FT of the intercorrelations (cross-correlation spectrum) between observed fringes at the reference (λ_{ref}) and considered (λ) wavelengths is given by,

$$
\begin{aligned}
C(u, v, u_{ref}, v_{ref}, \lambda, \lambda_{ref}) &= \langle V_{obs}(u, v, \lambda, t) V_{obs}^*(u_{ref}, v_{ref}, \lambda_{ref}, t) \rangle_t \\
&= |C(u, v, u_{ref}, v_{ref}, \lambda, \lambda_{ref})| e^{i\phi_{diff}(u,v,u_{ref},v_{ref},\lambda,\lambda_{ref})} ,
\end{aligned}
\tag{8.6}
$$

where ϕ_{diff} is the observed differential phase that can be approximated in a first order by

$$
\phi_{diff} = \phi - \left(a + \frac{b}{\lambda} \right) ,
\tag{8.7}
$$

where ϕ is defined in Eq. (8.1), the parameter a corresponds to a global phase offset, and b is a slope representing an overall residual piston term.

The term $(a + b/\lambda)$ that is subtracted from ϕ represents a weak dependence of the phase with λ. It contains not only the influence of the instrument and Earth atmosphere but also any phase term from the observed object that is slowly varying with λ. Consequently, only strong variations (orders higher than λ^{-1}) of the target's FT phase ϕ with wavelength remain in the differential phase ϕ_{diff}. Fortunately, this is typically what happens inside spectral lines where the phase can strongly depend on the wavelength, while it is nearly constant in the adjacent continuum, which is then often adopted as a reference wavelength range.

The *closure phase* is another observable commonly delivered by OLBI instruments operating simultaneously with three or more telescopes. From the fringes recorded with three telescopes, T_i, T_j, and T_k, the phases observed at each baseline are given by (c.f. Eq. (8.4))

$$
\begin{aligned}
\phi_{obs}(u_{ij}, v_{ij}) &= \phi(u_{ij}, v_{ij}) + \Delta\varphi(u_{ij}, v_{ij}) = \phi(u_{ij}, v_{ij}) + (\varphi_{T_i} - \varphi_{T_j}) , \\
\phi_{obs}(u_{jk}, v_{jk}) &= \phi(u_{jk}, v_{jk}) + \Delta\varphi(u_{jk}, v_{jk}) = \phi(u_{jk}, v_{jk}) + (\varphi_{T_j} - \varphi_{T_k}) , \\
\phi_{obs}(u_{ki}, v_{ki}) &= \phi(u_{ki}, v_{ki}) + \Delta\varphi(u_{ki}, v_{ki}) = \phi(u_{ki}, v_{ki}) + (\varphi_{T_k} - \varphi_{T_i}) ,
\end{aligned}
\tag{8.8}
$$

where only the dependence on the spatial frequencies are written explicitly for simplicity. The total phase shift $\Delta\varphi$ introduced by the *OTF* is explicitly represented in the last term as the combination of phase shifts (e.g. φ_{T_i}) introduced by the instrument and Earth atmosphere on the path of the light collected by each individual telescope (e.g. T_i). From the equation above it is possible to combine the three ϕ_{obs} to form the quantity called closure phase, in which the phase shifts φ are canceled out and only the sum of the phase information from the target at each baseline remains, i.e.,

$$
\begin{aligned}
\Phi_{ijk} &= \phi_{obs}(u_{ij}, v_{ij}) &+ \phi_{obs}(u_{jk}, v_{jk}) &+ \phi_{obs}(u_{ki}, v_{ki}) \\
&= \phi(u_{ij}, v_{ij}) &+ \phi(u_{jk}, v_{jk}) &+ \phi(u_{ki}, v_{ki})
\end{aligned}
\tag{8.9}
$$

The astrophysical results on fast-rotating stars presented in the following sections are based on measurements of visibility amplitudes, differential phases, and closure phases.

8.3 Roche-von Zeipel Model and the CHARRON Code

Most recent interferometric works on rapidly rotating, nondegenerate, single stars of intermediate to high masses adopt the Roche model (rigid rotation and mass concentrated in the center of the star) with a generalized form of the von Zeipel gravity darkening (von Zeipel 1924), hereafter called the RVZ model. Several existing codes provide numerical implementations of the RVZ model. The results presented here were obtained using the IDL-based program CHARRON (*code for high angular resolution of rotating objects in nature*). We present below a short description of the RVZ model and the CHARRON code. Rieutord (this proceedings) describes in more details the Roche model and in particular the gravity darkening effect on fast-rotating stars. A more detailed description of CHARRON is given by Domiciano de Souza et al. (2012a,b, 2002).

The stellar photospheric shape is assumed to follow the Roche equipotential (gravitational plus centrifugal),

$$\Psi(\theta) = -\frac{GM}{R(\theta)} - \frac{\Omega^2 R^2(\theta) \sin^2 \theta}{2} = -\frac{GM}{R_{eq}} - \frac{v_{eq}^2}{2} , \qquad (8.10)$$

where θ is the colatitude, G is the gravitation constant, M is the stellar mass, and R_{eq} and v_{eq} are the equatorial radius and rotation velocity. The last equality in the above equation sets the equipotential value from equatorial quantities. Solving this cubic equation provides the colatitude-dependent stellar radius $R(\theta)$.

The effective surface gravity is obtained from the gradient of Ψ calculated at the stellar surface $R(\theta)$,

$$g_{eff}(\theta) = |\mathbf{g}_{eff}(R(\theta))| = |-\nabla \Psi| . \qquad (8.11)$$

In particular, the equatorial and polar effective gravity are given by

$$g_{eq} = g_{eff}(\frac{\pi}{2}) = \frac{GM}{R_{eq}^2} - \frac{v_{eq}^2}{R_{eq}} , \qquad (8.12)$$

and

$$g_p = g_{eff}(0) = g_{eff}(\pi) = \frac{GM}{R_p^2} . \qquad (8.13)$$

The ratio between the equatorial and polar radii can be obtained directly from Eq. (8.10),

$$\frac{R_{\text{eq}}}{R_{\text{p}}} = \left(1 - \frac{v_{\text{eq}}^2 R_{\text{p}}}{2GM}\right)^{-1} = 1 + \frac{v_{\text{eq}}^2 R_{\text{eq}}}{2GM}$$

or (8.14)

$$\epsilon \equiv 1 - \frac{R_{\text{p}}}{R_{\text{eq}}} = \frac{v_{\text{eq}}^2 R_{\text{p}}}{2GM},$$

where ϵ is the flattening parameter, ranging from 0 (spherical star where $r = R_{\text{p}}$ at all latitudes) to a maximum (critical) flattening ϵ_c, attained when gravity is totally compensated by the centrifugal force at some point on the stellar surface. This condition ($g_{\text{eff}} = 0$) is satisfied first at the stellar equator where the centrifugal force is highest, with the equatorial radius attaining its critical (maximum) value $R_c = 3/2R_{\text{p}}$, so that $\epsilon_c = 1/3$. By imposing that $g_{\text{eq}} = 0$ (Eq. (8.12)) and solving for the equatorial velocity, the critical equatorial rotation velocity v_c and the critical angular rotation velocity Ω_c in the Roche model can be defined as

$$v_c = \sqrt{\frac{GM}{R_c}},$$

and (8.15)

$$\Omega_c = \frac{v_c}{R_c} = \sqrt{\frac{GM}{R_c^3}}.$$

Gravity darkening is considered by relating the local effective gravity $g_{\text{eff}}(\theta)$ ($= |\nabla\Psi(\theta)|$) to the local effective temperature $T_{\text{eff}}(\theta)$ (and local radiative flux $F(\theta)$) by,

$$T_{\text{eff}}(\theta) = \left(\frac{F(\theta)}{\sigma}\right)^{0.25} = \left(\frac{C}{\sigma}\right)^{0.25} g_{\text{eff}}^\beta(\theta),$$ (8.16)

where σ is the Stefan–Boltzmann constant and β is the gravity-darkening coefficient, which is more general than the value from von Zeipel (1924): $\beta = 0.25$. However, β is still assumed to be constant over the stellar surface (see Rieutord, this proceedings, for a more detailed discussion on gravity darkening). The proportionality constant C can be related to the stellar luminosity L and the average effective temperature $\overline{T}_{\text{eff}}$ over the total stellar surface $S_\star$,

$$L = \sigma \int T_{\text{eff}}^4(\theta)\, dS = \sigma \overline{T}_{\text{eff}}^4 S_\star = C \int g_{\text{eff}}^{4\beta}(\theta)\, dS.$$ (8.17)

In our numerical implementation of the RVZ model (CHARRON code) the stellar surface is divided into a predefined grid with nearly identical surface area elements (typically $\sim$50,000 surface elements). From $T_{\mathrm{eff}}(\theta)$ and $g_{\mathrm{eff}}(\theta)$ defined in the equations above, a local specific intensity from a plane-parallel atmosphere $I = I(g_{\mathrm{eff}}, T_{\mathrm{eff}}, \lambda, \mu)$ is associated to each surface element. Here, λ is the wavelength and μ is the cosine between the normal to the surface grid element and the line of sight (limb darkening is thus automatically included in the model). The local specific intensities I are interpolated from a grid of specific intensities, which are pre-calculated using spectral synthesis codes and model atmospheres available in the literature.

From the juxtaposition of specific intensities associated to each surface element we obtain wavelength-dependent intensity maps of the visible stellar surface at the chosen spectral domain and resolution, such as the images given in Fig. 8.1. The interferometric observables (e.g., squared visibilities, closure phases, differential phases) are then directly obtained from the Fourier transform of these sky-projected photospheric intensity maps, which for a given star in the sky also depend on its rotation-axis inclination angle i and on the position angle of its sky-projected rotation axis $\mathrm{PA}_{\mathrm{rot}}$ (counted from north to east until the visible stellar pole).

Indeed, the geometrical flattening and gravity darkening caused by rotation shown in Fig. 8.1 can be measured by OLBI (Sect. 8.4). In particular, the small differences of intensity distributions among different wavelengths in this figure can be measured by the spectro-interferometric techniques described in Sect. 8.2.2. The darker vertical lines corresponding to the location of the photospheric absorption lines at different wavelengths (center and right panels) introduce small shifts of interferometric fringe positions (relative to the continuum) that lead to a "S"-shaped

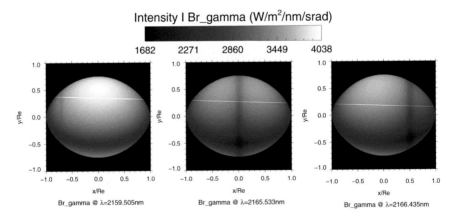

Fig. 8.1 Specific intensity I_{λ} in the vicinity of the hydrogen spectral line Brγ: continuum (*left panel*), line center (*middle panel*), and line center plus Doppler shift of $0.5 v_{\mathrm{eq}} \sin i$ (*right panel*). These intensity maps were created with the model CHARRON for a fast-rotating B type star with $M = 6 M_{\odot}$, $R_{\mathrm{eq}} = 11 R_{\odot}$, $\overline{T}_{\mathrm{eff}} = 15{,}000$ K, $\beta = 0.2$, $i = 60°$, and $v_{\mathrm{eq}} = 290$ km/s $= 0.929 v_{\mathrm{c}}$. The signatures of rotation seen on these intensity maps can be measured by OLBI techniques

signal of rotation. This signal can be detected in the differential phases observed across photospheric lines (Sect. 8.5 and Fig. 8.4).

8.4   Gravity Darkening of Achernar Measured from OLBI at ESO-VLTI

In this section we analyze interferometric observations of the Be star Achernar recorded at the ESO-VLTI (Haguenauer et al. 2010) to measure several stellar parameters (size, inclination, rotation velocity), and in particular the gravity darkening induced by rotation. Further details of this analysis are given by Domiciano de Souza et al. (2014).

Achernar is a Be star alternating between periods where it behaves mostly as a normal B star and periods with emission lines from a circumstellar disk in a time scale of typically 10 years. Thus, before considering Achernar as a simple fast-rotator without disk in our analysis, we performed a physical modeling of its close circumstellar environment (CSE) to investigate multi-technique observations (spectroscopic, polarimetric, and photometric) in order to show that the VLTI data were recorded during a normal B phase. Our analysis shows that any possible influence of a residual disk can be neglected within $\sim \pm 1\,\%$ level of intensity. This conclusion is also supported by interferometric image reconstruction.

Near-infrared (H band) interferometric data of Achernar were obtained in 2011/Aug.-Sep. and 2012/Sep. with the VLTI/PIONIER beam combiner (Le Bouquin et al. 2011). We used the largest quadruplet available with the Auxiliary Telescopes (AT) at that time (AT stations A1-G1-K0-I1) to resolve the stellar photosphere as much as possible. The resulting uv coverage (Fourier plane coverage) is quite satisfactory as shown in Fig. 8.2. Data were reduced and calibrated with the package pndrs (Le Bouquin et al. 2011). Each observation provides six squared visibilities V^2 and four closure phases CP.

The combination of (1) high quality interferometric observations of Achernar taken in a normal B phase, (2) physical modeling of fast rotators, and (3) model fitting with an efficient MCMC (Markov chain Monte Carlo) method (using the emcee code; Foreman-Mackey et al. 2013), allowed a robust and precise determination of photospheric parameters for this Be star. The results of this analysis is summarized in Table 8.1, which gives the measured physical parameters of Achernar and their corresponding uncertainties. Derived parameters based on measured values are also listed in the table. The reduced χ^2 of the best-fit model is $\chi_r^2 = 1.9$, where χ^2 has its usual definition and is composed by the sum of χ^2 from the V^2 and CP data. The sky-projected intensity map in the H band of the visible stellar photosphere for this best-fit model is shown in Fig. 8.3. This best-fit model is also compatible with polarimetric, photometric, spectroscopic observations of Achernar (further details are given by Domiciano de Souza et al. 2014).

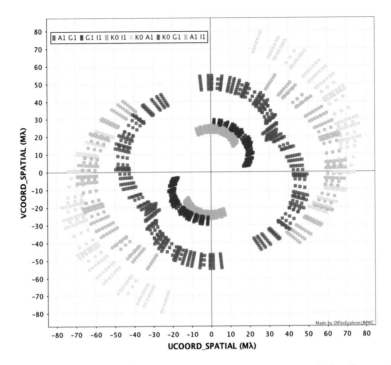

Fig. 8.2 *uv* coverage of VLTI/PIONIER observations of Achernar. The AT baselines used are identified with different colors. Image adapted from the OIFITSExplorer/JMMC tool

Being the most flattened fast-rotating, non-degenerated, single star known to-date, Achernar provides a crucial test to gravity-darkening theories. We confronted the results obtained from the analysis of PIONIER data described above to the gravity-darkening model proposed by Espinosa Lara and Rieutord (2011, ELR model; see also Rieutord, these proceedings). Figure 5 of Rieutord (these proceedings) or Fig. 13 of Domiciano de Souza et al. (2014) compares the gravity darkening β coefficient determined from the MCMC fitting of the CHARRON model to the VLTI/PIONIER data, as well as with values from other fast rotators (from other authors), with an equivalent β computed from the ELR model. The gravity darkening coefficient measured on Achernar is compatible with the previsions of the ELR model providing observational support to this model in the regime of highly flattened stars (flattening $\epsilon = 1 - R_{\mathrm{p}}/R_{\mathrm{eq}} > 0.26$ or, equivalently, $R_{\mathrm{eq}}/R_{\mathrm{p}} > 1.35$).

Table 8.1 Physical parameters and uncertainties of Achernar derived from the fit of the RVZ model (CHARRON code) to VLTI/PIONIER H-band data (visibility amplitudes and closure phases) using the MCMC method (emcee code; Foreman-Mackey et al. 2013)

Fitted model parameters	Values and uncertainties
Equatorial radius: R_{eq} $(R_\odot)^a$	9.16 (+0.23; −0.23)
Equatorial rotation velocity: v_{eq} (km/s)	298.8 (+6.9; −5.5)
Rotation-axis inclination angle: i (°)	60.6 (+7.1; −3.9)
Gravity-darkening coefficient: β	0.166 (+0.012; −0.010)
Position angle of the visible pole: PA_{rot} (°)	216.9 (+0.4; −0.4)
Derived model parameters	Values
Equatorial angular diameter: $\oslash_{eq} = 2R_{eq}/d$ (mas)^b	1.99
Polar radius: R_p $(R_\odot)$	6.78
R_{eq}/R_p; $\epsilon \equiv 1 - R_p/R_{eq}$	1.352; 0.260
$v_{eq} \sin i$ (km/s)	260.3
Critical rotation rate: v_{eq}/v_c	0.883
Polar and equatorial temperatures: T_p (K) ; T_{eq} (K)	17 124; 12 673
Luminosity: $\log L/L_\odot$	3.480

The minimum reduced χ^2 of the best-fit model is $\chi_r^2 = 1.9$ (for 1777 degrees of freedom and 5 free parameters). The fixed parameters of the model are the stellar mass $M = 6.1\,M_\odot$, surface averaged effective temperature $\overline{T}_{eff} = 15{,}000$ K, and distance $d = 42.75$ pc (van Leeuwen 2007). The stellar parameters derived from the best-fit RVZ model are also listed. It is important to note that this best-fit model is also compatible with polarimetric, photometric, spectroscopic observations of Achernar. Details of this analysis, discussion of the results, and further references are given by Domiciano de Souza et al. (2014)
^{a}The uncertainty in the distance d from van Leeuwen (2007) was added quadratically to the fit-uncertainty on R_{eq}
^{b}mas stands for milliarcseconds

The ELR model thus seems validated by interferometric results obtained on five hot, massive fast-rotating stars, with flattening ϵ ranging from 0.11 to 0.26 (R_{eq}/R_p from 1.13 to 1.35), Vega (α Lyr) being the less flattened star, and Achernar being the flattest one. This agreement between theory and interferometric observations provides a more realistic description of gravity darkening on single stars, significantly improving our view of this important physical effect since the pioneering work of von Zeipel almost a century ago. This more profound understanding of gravity darkening provided by the ELR model also allows to decrease the number of parameters required to model fast rotators, since the surface intensity (effective temperature) distribution is defined without the need of a β coefficient, present in the von Zeipel-like gravity darkening laws currently used.

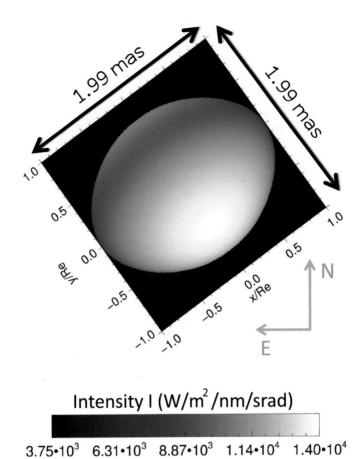

Intensity I (W/m²/nm/srad)

$3.75 \cdot 10^3$ $6.31 \cdot 10^3$ $8.87 \cdot 10^3$ $1.14 \cdot 10^4$ $1.40 \cdot 10^4$

Fig. 8.3 Modeled H-band intensity map of Achernar corresponding to the best MCMC fit of the CHARRON RVZ model to the VLTI/PIONIER observations. The spatial coordinates are given in angular milliarcseconds (mas) units and also normalized to the equatorial radius $R_{eq} = 9.16 R_\odot$. The complete list of the measured stellar parameters is given in Table 8.1

8.5 Beyond the Diffraction Limit of OLBI

The results described in the previous section providing several physical parameters of a fast-rotating star can be obtained only for stars that can be well resolved (spatially) by the interferometer, i.e., for stars presenting angular diameters of the same order of the classically adopted diffraction-limited angular resolution: λ / B_{proj}^{max}, where λ is the wavelength and B_{proj}^{max} is the maximum projected baseline. With available baselines B_{proj}^{max} of $\simeq 100$–300 m the current optical/IR interferometers can resolve angular diameters of ≤ 0.5–2 mas, which are typical values for many bright stars.

For stars with sizes several times ($\sim$5–10) smaller than $\lambda/B_{\text{proj}}^{\text{max}}$ the analysis applied in the previous section cannot be performed and new/alternative high angular resolution techniques are required to map the surface of fast-rotating stars. Differential interferometry is a self-calibrated and essentially seeing-independent technique that combines high spatial and high spectral resolutions and allows to go beyond diffraction-limited angular resolution of the interferometer. This allows differential interferometry to be used even in cases where visibility amplitudes $|V|$ (fringe contrast) are unavailable and/or when the star is poorly resolved (a few times smaller than the diffraction limit of the instrument) so that $|V| \simeq 1$. The differential phase described in Sect. 8.2.2 is a commonly used differential-interferometry observable that allows to measure sizes, rotation velocities, and orientation of rotating stars.

This possibility was theoretically proposed by Chelli and Petrov (2005) and was recently demonstrated on four fast-rotating stars based only on differential phase measurements (Domiciano de Souza et al. 2012a; Hadjara et al. 2014). As an example, we present below the results obtained on the Be star Achernar based on differential phases recorded with the ESO VLTI/AMBER beam combiner (Petrov et al. 2007) centered on the Brγ hydrogen line (wavelengths ranging from 2.159 to 2.172 µm).

VLTI/AMBER observations of Achernar in high spectral resolution ($\lambda/\Delta\lambda \approx$ 12,000) on the K band were carried out from October 25th to November 1th, 2009, during four nights with a different Auxiliary Telescope (AT) triplet configuration in each night, providing good (u, v) coverage (as complete as in Fig. 8.2). After a tricky data reduction, described in detail by Domiciano de Souza et al. (2012a), our final data set consists of 84 ($= 28 \times 3$ baselines) $\phi_{\text{diff}}(\lambda)$ curves centered on Brγ and presents $\simeq$45 ϕ_{diff} points for each of the 84 individual projected baselines.

These VLTI/AMBER ϕ_{diff} observations of Achernar were analyzed with the numerical model CHARRON presented in Sect. 8.3 used to perform a χ^2 minimization with an IDL implementation of the Levenberg–Marquardt (LM) algorithm (Markwardt 2009, and references therein). The physical parameters of Achernar derived from this analysis are summarized in Table 8.2. Some examples of observed and modeled ϕ_{diff} associated to the best-fit model are shown in Fig. 8.4.

This result shows that differential interferometry can potentially be used to measure stellar parameters of fast-rotators, namely R_{eq} (or $\oslash_{\text{eq}}$), v_{eq}, i, and PA$_{\text{rot}}$. The derived $\oslash_{\text{eq}}$ is $\simeq$1.5 times smaller than the maximum available diffraction-limited angular resolution. Indeed, these observations show that the Achernar's diameter is already resolved at shorter baselines, with a clear signature of rotation (of the order of the noise level of ϕ_{diff}) already seen at B_{proj} as short as $\simeq$30 m (Fig. 8.4). For example, considering the observations with $B_{\text{proj}} \simeq 45$ m, the measured diameter is $\simeq$4 times smaller than the corresponding diffraction-limited

Table 8.2 Parameters and uncertainties estimated from a Levenberg-Marquardt fit of our model to the VLTI/AMBER ϕ_{diff} observed on Achernar

Fitted model parameters	Values and uncertainties
R_{eq} ($R_{\odot}$)[a]	11.6 (+0.4; −0.4)
v_{eq} (km/s)	298 (+9; −9)
i (°)	78.5 (+5.2; −5.2)
PA_{rot} (°)	214.9 (+1.6; −1.6)
Derived model parameters	Values
$\oslash_{\mathrm{eq}}$ (mas)	2.45
R_{p} ($R_{\odot}$)	8.0
$R_{\mathrm{eq}}/R_{\mathrm{p}}$; $\epsilon \equiv 1 - R_{\mathrm{p}}/R_{\mathrm{eq}}$	1.45; 0.31
$v_{\mathrm{eq}} \sin i$ (km/s)	292
$v_{\mathrm{eq}}/v_{\mathrm{c}}$	0.96
T_{p} (K) ; T_{eq} (K)	18 013; 9 955
$\log L/L_{\odot}$	3.654

The minimum reduced χ^2 of the fit is $\chi_{\mathrm{r}}^2 = 1.2$ (for 3809 degrees of freedom and 4 free parameters). The fixed parameters of the model-fitting are the stellar mass $M = 6.1 M_{\odot}$, surface averaged effective temperature $\overline{T}_{\mathrm{eff}} = 15,000$ K, gravity-darkening coefficient $\beta = 0.2$, and distance $d = 44.1$ pc (Perryman et al. 1997)
[a] The uncertainty in the distance d from Perryman et al. (1997) was added quadratically to the fit-uncertainty on R_{eq}

resolution of 10 mas for this baseline length, revealing the super-resolution capacity of this technique.

Although differential interferometry is a promising technique to measure stellar parameters of fast-rotators, the comparison of these results with those obtained from classical interferometric observables (Table 8.1) shows that some improvements are still necessary. For example, the model-fitting of differential phases can a present relatively shallow χ^2 because the data points in the continuum (equal to zero by construction; see Fig. 8.4) can always be well fitted by any set o model parameters. Also, compared to visibility amplitudes and closure phases, the differential phases are also less sensitive to the β value and thus less capable of disentangling i and v_{eq} (see also Domiciano de Souza et al. 2012a). Possibilities to overcome these difficulties in order to optimally exploit the differential phases are for example to adopt model-fitting procedure that allow a robust and realistic estimation of the uncertainties (as the MCMC method for example), and/or to combine the differential phase observations with additional observational constraints, such as spectroscopy, polarimetry, and photometry. We follow this strategy in the next section where we analyze both the PIONIER and AMBER data together using the MCMC method.

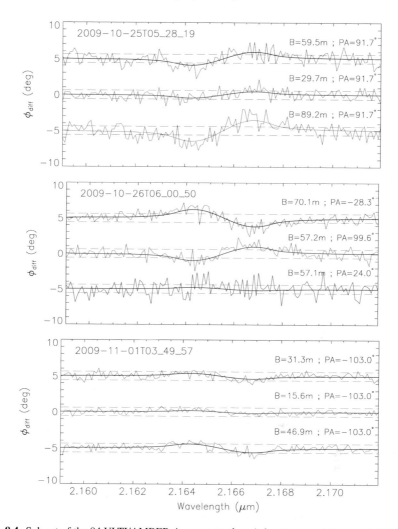

Fig. 8.4 Sub-set of the 84 VLTI/AMBER ϕ_{diff} measured on Achernar around Br γ at 28 different observing times (format YYYY-MM-DDTHH_MM_SS) and, for each time, three different projected baselines B and baseline position angles PA, as indicated in the plots. The *dashed gray horizontal lines* indicate the median ϕ_{diff} uncertainty $\pm 0.6°$ of all observations. The *smooth curves* superposed to the observations are the best-fit ϕ_{diff} obtained with a χ^2 minimization using the CHARRON code. All *curves* have zero average value in the continuum, but they were shifted for better readability. The full data set and detailed description of the data analysis is given by Domiciano de Souza et al. (2012a)

8.6 Combined Analysis of AMBER and PIONIER Data

In this section we analyze the PIONIER H-band data (visibility amplitudes and closure phases) and AMBER K-band data (differential phases centered on Br γ) together using the MCMC method. We applied the same model-fitting strategy

Table 8.3 Physical
parameters and uncertainties
of Achernar derived from the
fit of the RVZ model
(CHARRON code) to both
PIONIER H-band data
(visibility amplitudes and
closure phases) and AMBER
K-band data (differential
phases centered at Br γ) using
the MCMC method (emcee
code; Foreman-Mackey et al.
2013)

Fitted model parameters	Values and uncertainties
R_{eq} $(R_\odot)^a$	9.22 (+0.25; −0.24)
v_{eq} (km/s)	306.0 (+10.2; −10.4)
i (°)	59.0 (+12.7; −6.6)
β	0.186 (+0.011; −0.010)
PA_{rot} (°)	216.6 (+0.4; −0.4)
Derived model parameters	Values
$\varnothing_{eq}$ (mas)	2.01
R_p $(R_\odot)$	6.72
R_{eq}/R_p; $\epsilon \equiv 1 - R_p/R_{eq}$	1.371; 0.271
$v_{eq} \sin i$ (km/s)	262.4
v_{eq}/v_c	0.861
T_p (K) ; T_{eq} (K)	17 467; 12 082
$\log L/L_\odot$	3.479

The minimum reduced χ^2 of the fit is $\chi_r^2 = 1.5$
(5590 degrees of freedom). The fixed parameters of
the model are the stellar mass $M = 6.1 M_\odot$, surface
averaged effective temperature $\overline{T}_{eff} = 15,000$ K, and
distance $d = 42.75$ pc (van Leeuwen 2007). This best-fit
model the combination of PIONIER and AMBER data
is compatible with the results in Table 8.1. This solution
is also compatible with spectroscopic, polarimetric, and
photometric observations
[a]The uncertainty in the distance d from van Leeuwen
(2007) was added quadratically to the fit-uncertainty
on R_{eq}

described in Sect. 8.4. The results of the MCMC model-fitting using the CHARRON
code (RVZ model) are presented in Table 8.3.

The stellar parameters values shown in Table 8.3 are compatible with the values
in Table 8.1 (within $\simeq \pm 1\sigma$). We note however, that the uncertainty on v_{eq}
and i is increased when the AMBER differential phases are included, since these
observables cannot easily disentangle those two quantities.

This analysis shows that a unique solution can be obtained from the whole
set of interferometric data obtained in a normal B star phase (AMBER and
PIONIER). The best-model obtained from these two data sets are also compatible
with spectroscopic, polarimetric, and photometric observations since they agree
with the results from Table 8.1.

Acknowledgements I am grateful to the organizers of the Besançon school for their invitation
to write this paper. PIONIER is funded by the Université Joseph Fourier (UJF), the Institut de
Planétologie et d'Astrophysique de Grenoble (IPAG), the Agence Nationale pour la Recherche
(ANR-06-BLAN-0421 and ANR-10-BLAN-0505), and the Institut National des Science de
l'Univers (INSU PNP and PNPS). The integrated optics beam combiner is the result of a
collaboration between IPAG and CEA-LETI based on CNES R&T funding. This research has made
use of the SIMBAD database, operated at the CDS, Strasbourg, France, of NASA Astrophysics

Data System Abstract Service.[2] We also have used the Jean-Marie Mariotti Center (JMMC) services `OIFits Explorer`,[3] and `SearchCal`.[4]

References

Born, M., & Wolf, E. (1999). *Principles of optics: Electromagnetic theory of propagation, interference and diffraction of light* (7th ed.). Cambridge University Press.

Chelli, A., & Petrov, R. G. (2005). Model fitting and error analysis for differential interferometry. II. Application to rotating stars and binary systems. *Astronomy & Astrophysics, Supplement, 109*, 401.

Domiciano de Souza, A., Hadjara, M., Vakili, F., Bendjoya, P., Millour, F., Abe, L., et al. (2012a). Beyond the diffraction limit of optical/IR interferometers. I. Angular diameter and rotation parameters of Achernar from differential phases. *Astronomy & Astrophysics, 545*, A130.

Domiciano de Souza, A., Kervella, P., Jankov, S., Abe, L., Vakili, F., di Folco, E., et al., (2003). The spinning-top Be star Achernar from VLTI-VINCI. *Astronomy & Astrophysics, 407*, L47.

Domiciano de Souza, A., Kervella, P., Moser Faes, D., Dalla Vedova, G., Mérand, A., Le Bouquin, J.-B., et al. (2014). The environment of the fast rotating star Achernar. III. Photospheric parameters revealed by the VLTI. *Astronomy & Astrophysics, 569*, A10.

Domiciano de Souza, A., Vakili, F., Jankov, S., Janot-Pacheco, E., & Abe, L. (2002). Modelling rapid rotators for stellar interferometry. *Astronomy & Astrophysics, 393*, 345.

Domiciano de Souza, A., Zorec, J., & Vakili, F. (2012b). CHARRON: Code for high angular resolution of rotating objects in nature. In S. Boissier, P. de Laverny, N. Nardetto, R. Samadi, D. Valls-Gabaud, & H. Wozniak (Eds.), *Proceedings of the Annual meeting of the French Society of Astronomy and Astrophysics* (pp. 321–324).

Espinosa Lara, F., & Rieutord, M. (2011). Gravity darkening in rotating stars. *Astronomy & Astrophysics, 533*, A43.

Foreman-Mackey, D., Hogg, D.W., Lang, D., & Goodman, J. (2013). emcee: The MCMC hammer. *Publications of the Astronomical Society of the Pacific, 125*, 306.

Hadjara, M., Domiciano de Souza, A., Vakili, F., Jankov, S., Millour, F., Meilland, A., et al. (2014). Beyond the diffraction limit of optical/IR interferometers. II. Stellar parameters of rotating stars from differential phases. *Astronomy & Astrophysics, 569*, A45

Haguenauer, P., Alonso, J., Bourget, P., Brillant, S., Gitton, P., Guisard, S., et al. (2010). The very large telescope Interferometer: 2010 edition. *Society of Photo-Optical Instrumentation Engineers (SPIE) Conference Series* (Vol. 7734).

Kervella, P., & Domiciano de Souza, A. (2006). The polar wind of the fast rotating Be star Achernar. VINCI/VLTI interferometric observations of an elongated polar envelope. *Astronomy & Astrophysics, 453*, 1059.

Le Bouquin, J.-B., Berger, J.-P., Lazareff, B., Zins, G., Haguenauer, P., Jocou, L., et al. (2011). PIONIER: A 4-telescope visitor instrument at VLTI. *Astronomy & Astrophysics, 535*, A67.

Markwardt, C. B. (2009). Non-linear least-squares fitting in IDL with MPFIT. In D. A. Bohlender, D. Durand, & P. Dowler (Eds.), *Astronomical Society of the Pacific Conference Series* (Vol. 411, pp. 251–254).

Millour, F., Meilland, A., Chesneau, O., Stee, Ph., Kanaan, S., Petrov, R., et al. (2011). Imaging the spinning gas and dust in the disc around the supergiant A[e] star HD 62623. *Astronomy & Astrophysics, 526*, A107.

---

[2] Available at http://cdsweb.u-strasbg.fr/.

[3] Available at http://www.jmmc.fr/oifitsexplorer.

[4] Available at http://www.jmmc.fr/searchcal.

Millour, F., Vannier, M., Petrov, R. G., Chesneau, O., Dessart, L., Stee, P., et al. (2006). Differential interferometry with the AMBER/VLTI instrument: Description, performances and illustration. In M. Carbillet, A. Ferrari, & C. Aime (Eds.), Astronomy with high contrast imaging III: Instrumental techniques, modeling and data processing. EAS Publications Series (Vol. 22, pp. 379–388). Cambridge: Cambridge University Press.

Perryman, M. A. C., Lindegren, L., Kovalevsky, J., Hoeg, E., Bastian, U., Bernacca, P. L., et al. (1997). The HIPPARCOS catalogue. *Astronomy & Astrophysics, 323*, L49.

Petrov, R. G., Malbet, F., Weigelt, G., Antonelli, P., Beckmann, U., Bresson, Y., et al. (2007). AMBER, the near-infrared spectro-interferometric three-telescope VLTI instrument. *Astronomy & Astrophysics, 464*, 1.

van Belle, G. T. (2012). Interferometric observations of rapidly rotating stars. *Astronomy & Astrophysics, Reviews, 20*, 51.

van Leeuwen, F. (2007). Validation of the new Hipparcos reduction. *Astronomy & Astrophysics, 474*, 653.

von Zeipel, H. (1924). The radiative equilibrium of a rotating system of gaseous masses. *Monthly Notices of the Royal Astronomical Society, 84*, 665.

Chapter 9
Doppler and Zeeman Doppler Imaging of Stars

Oleg Kochukhov

Abstract In this chapter we discuss the problem of reconstructing two-dimensional stellar surface maps from the variability of intensity and/or polarisation profiles of spectral lines. We start by outlining the main principles of the scalar Doppler imaging problem concerned with recovering maps of chemical spots, temperature or brightness from the intensity spectra. After presenting the physical and mathematical foundations of this remote sensing method, we review its applications to mapping different types of spots in early-type chemically peculiar and late-type active stars, and non-radial pulsations in early-type stars. We also discuss an extension of Doppler imaging to the problem of recovering vector distributions of stellar magnetic fields from spectropolarimetric observations and review applications of this Zeeman Doppler imaging technique in the context of stellar magnetism studies.

9.1 Introduction

Stars exhibit different types of inhomogeneities on their surfaces. In many cases, including the presence of cool spots on the solar surface, magnetic fields are responsible for this structure formation. In other situations lateral inhomogeneities may be related to non-radial pulsations or surface convection. Investigation of the formation, evolution and mutual interaction of different stellar surface structures represents an essential part of stellar physics, which has profound consequences for understanding the stellar evolution in general and the phenomena of mass loss, angular momentum evolution, planet formation and habitability of exoplanets in particular. Detailed information about geometrical distribution of stellar surface inhomogeneities and its temporal evolution comprises a critical input required for developing realistic theoretical models of these phenomena. For instance, historical observations of sunspot cycles and more recent discovery of the link between cool spots and magnetic fields was a prerequisite for understanding the solar activity in the context of a dynamo theory.

O. Kochukhov (✉)

Department of Physics and Astronomy, Uppsala University, Box 516, SE 75120 Uppsala, Sweden
e-mail: oleg.kochukhov@physics.uu.se

© Springer International Publishing Switzerland 2016   177
J.-P. Rozelot, C. Neiner (eds.), *Cartography of the Sun and the Stars*,
Lecture Notes in Physics 914, DOI 10.1007/978-3-319-24151-7_9

For the Sun a bewildering complexity of surface structures is accessible to direct imaging with many ground-based instruments and space missions. But, with a few exceptions of interferometric studies of nearby and intrinsically large stars (see chapters by P. Kervella and J. Monnier), the disks of stars other than the Sun cannot be resolved and hence cannot be studied using direct imaging. In that case the only viable option to obtain information about the stellar surface structures is to apply some form of an inverse remote sensing method of *indirect imaging*, capable of recovering a stellar surface map from spatially unresolved stellar observations.

In this chapter we will discuss two powerful remote sensing techniques: Doppler imaging (DI) and Zeeman Doppler imaging (ZDI). The first method uses time series observations of the intensity line profiles recorded at high spectroscopic resolution to reconstruct scalar star spot maps. The second technique utilises circular polarisation signatures inside line profiles or, more generally, high-resolution spectropolarimetric observations in all four Stokes parameters to recover topology of vector magnetic field at the stellar surface. In the following sections we outline key physical principles and mathematical methods used by DI and ZDI techniques. This discussion is supplemented with a brief overview of the applications of indirect imaging to different types of surface structures in both hot and cool stars.

9.2 Doppler Imaging with Intensity Spectra

9.2.1 Main Principles of DI

An inhomogeneity on the stellar surface changes the local emergent line and continuum radiation. For example, a local enhancement of the concentration of some chemical element increases intensity of its absorption lines; the presence of a cool dark spot changes the local spectrum and suppresses the continuum radiation relative to immaculate photosphere. The flux spectrum of an unresolved stellar disk represents a weighted average of all local spectral contributions, Doppler-shifted according to the local projected rotational velocity relative to the observer. If the surface inhomogeneities are sufficiently large and have a high contrast, their spectral signatures will be visible in the disk-integrated line profiles as distortions—either "emission" bumps for cool spots or an extra absorption features for spots with an enhanced element concentration (see Fig. 9.1a, b)—superimposed onto the regular Doppler-broadened line profile. The velocity of the distortion relative to the line centre is given by the longitude of the surface feature, reckoned from the central stellar meridian.

The latitude position of a star spot cannot be inferred from a single observation. Instead, one uses time-series behaviour of the spot signatures to recover the latitude information. As the star rotates, spectral distortions corresponding to each spot first appear on the blue side of the spectral line profile and then gradually move to the red side. As illustrated by the dynamic spectra in Fig. 9.1c, temporal behaviour of

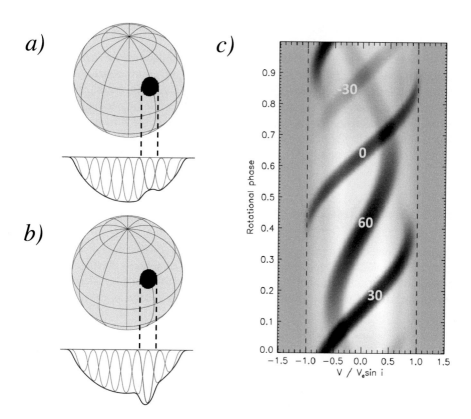

Fig. 9.1 Illustration of the main principles of Doppler imaging. The local stellar spectrum formed in a spot differs from the photospheric spectrum. This leads to a distortion in the disk-integrated stellar line profile. This distortion is Doppler-shifted according to the position of the spot relative to the disk centre. The two spherical plots and corresponding line profiles illustrate spectral signatures of (**a**) a single spot with reduced continuum brightness typical of late-type active stars and (**b**) a spot with enhanced line strength typical of early-type chemically peculiar stars. As the star rotates, the spot signature moves across the line profile from *blue* to *red*. The *rectangular panel* (**c**) shows the dynamic difference spectrum as a function of the rotational phase. In this case, the stellar surface has four small spots at latitudes −30, 0, +30, and +60 degrees. This plot demonstrates how temporal variation of the spot signatures depends on their latitude position

the spot signatures depends significantly on their latitudes. For the intermediate inclination of the stellar rotational axis ($i = 50°$) shown in this figure, the line profile distortions produced by the spots with latitudes from $0°$ to $\approx +30°$ span the largest range of relative velocities and are visible roughly half of the rotational period. On the other hand, signatures of structures at higher latitudes move over a smaller velocity range and are visible over a larger fraction of the rotational cycle. Spots with a latitude above i remain visible all the time; their line profile signatures exhibit faint "backward" red-to-blue progression during some part of the rotational cycle. Spots below the equator are visible very briefly and leave a weak signature in the disk-integrated line profile.

To summarise, the amplitude and position of stellar surface structures are encoded in the variability of distortions observed in Doppler-broadened line profiles. Given high-quality observations of stellar spectra, obtained with a sufficiently dense rotational phase coverage and a signal-to-noise (S/N) ratio high enough to clearly detect line profile distortions due to spots, it should be possible to reconstruct a two-dimensional map of the stellar surface. This is the scope of the inversion technique known as Doppler imaging.

9.2.2 Spatial Resolution of DI

A classical method of estimating resolution at the stellar surface provided by DI is to compare the rotational Doppler broadening with the stellar line width in the absence of rotation. The latter is usually dominated by the instrumental broadening. Then, the angular size of the resolution element at the stellar equator is given in degrees by

$$\delta\ell = 90° \frac{\Delta\lambda}{\lambda} \frac{v_c}{v_e \sin i}, \tag{9.1}$$

where $R = \lambda/\Delta\lambda$ is the instrumental resolution of the spectrograph, $v_e \sin i$ is the projected rotational velocity, and v_c is the speed of light. For example, for $R = 65\,000$ and $v_e \sin i = 150\,\mathrm{km\,s^{-1}}$ we obtain $\delta\ell = 2.7°$, which corresponds to 133 resolution element along the equator. The distance to the star is not directly entering Eq. (9.1); but of course the star has to be sufficiently bright to allow obtaining high S/N ratio spectra within a time interval small compared to the rotational period. Considering, for instance, a main sequence early B-type star at 1 kpc (with V magnitude ≈ 6 it is easily accessible to high-resolution spectroscopy at 2–4 m class telescopes), we can infer that $\delta\ell = 2.7°$ corresponds to an angular resolution of 0.3 μarcsec. This is orders of magnitude better than any direct imaging technique can provide at the moment or in the foreseeable future.

A common misconception is to use Eq. (9.1) for estimating the lower $v_e \sin i$ limit of DI, which leads to an argument that DI requires $v_e \sin i \gg v_c/R$. On the one hand, it is certainly true that the spatial resolution of this remote sensing method gradually decreases as the Doppler broadening becomes comparable to the intrinsic line width. On the other hand, we should not forget that Eq. (9.1) estimates spatial resolution based on a single *snapshot* observation. In reality, DI operates with *time series* spectra, gaining significant additional resolution from rotational modulation. The same type of information is used by the photometric mapping methods (see the chapter by A. Lanza) to recover coarse brightness maps of stellar surfaces from broad-band light curves. Similarly, DI and especially its extension to the polarisation spectra—ZDI—is capable of recovering maps of large-scale surface features for extremely slowly rotating stars (e.g. Petit et al. 2008) independently of the actual $v_e \sin i$ value, provided that the input time series data exhibit a significant phase-dependent variation.

9.2.3 DI as an Ill-Posed Inverse Problem

As discussed above, DI provides a method to relate the observed line profile variability to the underlying geometry of stellar surface structures. However, a key question is how to mathematically implement the *inverse problem* of deriving a 2-D surface map from a given spectroscopic observational data set. Early studies of chemically peculiar stars and late-type active stars attempted to fit observations with parameterised maps, typically consisting of a small number of circular spots (Mihalas 1973; Vogt and Penrod 1983b) or employing a low-order spherical harmonic expansion (Mégessier 1975). But it was quickly realised that the problem of finding parameters of these spots is in practice mathematically ill-posed, meaning that an infinite number of very different solutions can fit a given data set. A breakthrough was achieved when Goncharskii et al. (1977) suggested to employ regularisation methods to ensure uniqueness of the DI solution. In other words, besides observations themselves, one introduces some additional criterium of simplicity to limit the family of possible solutions. Mathematically this is implemented by finding a solution x that minimises the sum of χ^2 of the fit to observations and regularisation function R

$$\sum [S_{\text{obs}} - S_{\text{model}}(x)]^2 / \sigma_{\text{obs}}^2 + \Lambda R(x) \to \min, \qquad (9.2)$$

where regularisation parameter Λ is determined empirically by trial and error or is set by requiring a certain target χ^2.

In the early studies of chemical spots in Ap stars (e.g. Khokhlova et al. 1986) DI problems were regularised with the Tikhonov functional

$$R(x) = \|\nabla x\|. \qquad (9.3)$$

Later Vogt et al. (1987) introduced the maximum entropy method (MEM) in the context of the application of DI to temperature mapping of cool active stars. In this case,

$$R(x) = \sum_i \frac{x_i}{x_0} \log \frac{x_i}{x_0}, \qquad (9.4)$$

where x_0 is some default value of the map. Subsequently both regularisation methods were used for mapping chemical structures in Ap stars and temperature spots in late-type stars (Piskunov et al. 1990; Hatzes 1991).

The two regularisation approaches differ slightly in their interpretation of the concept of "simplicity" of a surface map. The Tikhonov regularisation favours solutions with the least local gradient. On the other hand, MEM prefers a map with the least deviation from the default value. In practice, the two methods produce compatible results when applied to the same observations of sufficiently high quality (Strassmeier et al. 1991). However, it cannot be taken for granted that the two regularisation schemes are always equivalent and equally applicable to any DI problem. In particular, MEM fails whenever the surface structure does not have a natural default value. For example, this problem is encountered in the ZDI reconstruction of large-scale multipolar magnetic fields.

Finally, it should be added that a number of auxiliary stellar parameters are required to perform a DI inversion. They include the stellar rotational period P_{rot}, inclination of the stellar rotational axis relative to the line of sight i, the projected rotational velocity $v_e \sin i$, and the radial velocity of the star V_{r}. Of these parameters, P_{rot} and V_{r} are usually known prior to a DI analysis. The projected rotational velocity can be accurately determined with the help of DI inversions themselves by finding a $v_e \sin i$ value that yields the best fit to observations. An incorrect $v_e \sin i$ produces characteristic axisymmetric artefact features in the surface maps (Vogt et al. 1987; Rice et al. 1989). On the other hand, DI inversions are relatively insensitive to the choice of i, with errors of 10–15° leading to negligible distortions of the surface map. The relation

$$\sin i = \frac{P_{\mathrm{rot}} v_e \sin i}{50.613 R_\star} \tag{9.5}$$

can be employed to constrain inclination if the stellar radius $R_\star$ is known. In this equation $v_e \sin i$ is measured in $\mathrm{km\,s^{-1}}$, P_{rot} in days, and $R_\star$ in solar units.

In addition to all these parameters specific to DI, one has to know parameters that determine the shape and strength of the local line profiles. Depending of the modelling approach, these parameters can comprise either a few numbers required to specify some analytical line profile function (e.g. a Gaussian profile) or a complete set of stellar atmospheric parameters and chemical abundances for the most realistic spectrum synthesis representation of stellar observations.

9.2.4 DI Applications

Chemical Mapping of Ap and HgMn Stars

Chemical mapping of the upper main sequence chemically peculiar A and B stars was the first application of Doppler imaging (Goncharskii et al. 1983; Khokhlova and Pavlova 1984; Khokhlova et al. 1986). These objects, comprising a small fraction of all A and B stars, are distinguished by a host of peculiarities including anomalously strong spectral lines of iron-peak and rare-earth elements, a strongly non-uniform (with contrasts of up to several orders of magnitude) vertical and horizontal chemical distributions, and the presence of strong (typically 1–10 kG), globally organised magnetic fields. These so-called magnetic Ap stars exhibit a well-defined periodic variability of line profiles, spectral energy distribution, and magnetic field strength. Since the seminal paper by Stibbs (1950), this behaviour is interpreted in the context of the *oblique rotator model*. This phenomenological framework attributes all types of stellar variability to the rotational modulation of the aspect angle at which the surface spot structure and magnetic field are seen by a distant observer. Consequently, Ap stars vary in a strictly periodic manner and do not exhibit any intrinsic evolution of their surface structure. These properties make them ideal DI targets.

Early abundance DI studies of Ap stars focused on mapping a small number of chemical elements (Khokhlova et al. 1986; Ryabchikova et al. 1996; Rice and Wehlau 1991; Hatzes 1991, 1997), trying to relate their distributions to the process of atomic diffusion in magnetic field thought to be responsible for the formation of horizontal chemical inhomogeneities (Michaud et al. 1981). Although some encouraging results were obtained, particularly for Si (Alecian and Vauclair 1981), these interpretation efforts were hampered by the lack of detailed information about magnetic field geometries of the target stars.

More recent abundance DI studies of Ap stars provided maps of up to a dozen chemical elements (Kochukhov et al. 2004; Lüftinger et al. 2010; Nesvacil et al. 2012), taking advantage of a wide wavelength coverage of modern echelle spectrometers. Some of these maps were reconstructed simultaneously, by modelling blends containing contributions of several elements. A typical example of such abundance DI maps is shown in Fig. 9.2.

The comprehensive multi-element DI studies uncovered a complex and diverse behaviour that did not agree with expectations of simple atomic diffusion models for a star with a dipolar-like magnetic field. In particular, only a few chemical elements, notably Li (Polosukhina et al. 1999) and O (Rice et al. 1997, 2004), showed a surface distribution with a systematic relationship to the underlying magnetic field topology. Other chemical elements, for example Ca, Fe, Cr, Ti, typically show little or no such relation. There is also a great deal of diversity in the surface distributions of the same elements in stars with very similar fundamental and magnetic parameters and in the surface patterns of elements with similar properties (e.g. different rare-earth elements) in the same stars. This non-systematic behaviour suggests that some hitherto unknown structure formation mechanism contributes to shaping of chemical maps of Ap stars.

A new domain of applications of abundance DI was opened by the discovery of chemical spots in late-B HgMn stars (Ryabchikova et al. 1999). These objects also belong to the group of chemically peculiar stars. But, unlike magnetic Ap stars, HgMn stars lack detectable large- or small-scale magnetic fields (Aurière et al. 2010; Bagnulo et al. 2012; Kochukhov et al. 2013). Despite this, several chemical elements which are most overabundant in these stars (Hg, Y, Pt) show subtle line profile variations indicative of low-contrast non-uniform chemical distributions (Kochukhov et al. 2005; Hubrig et al. 2006; Folsom et al. 2010).

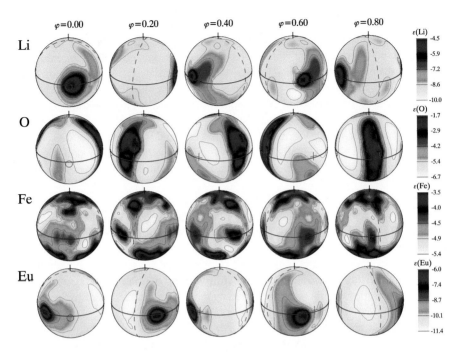

Fig. 9.2 Typical chemical abundance distributions reconstructed with DI for a magnetic Ap star. In this case, the maps of Li, O, Fe, and Eu are shown for the cool Ap star HD 83368. The scale bars to the right indicate abundance values in logarithmic units $\log(N_{el}/N_{tot})$. The magnetic field structure of this star is approximately dipolar, with a large magnetic obliquity ($\beta \approx 90°$). The "+" and "o" symbols indicate positions of the positive and negative magnetic poles, while the *dashed line* corresponds to the magnetic equator. Some chemical elements have simple distributions correlating with the dipolar magnetic field geometry. Other elements exhibit complex or simple maps, which show no apparent relation to the underlying magnetic field structure. Adapted from Kochukhov et al. (2004)

These inhomogeneities were mapped with DI in several HgMn stars (Adelman et al. 2002; Briquet et al. 2010; Makaganiuk et al. 2011, 2012; Korhonen et al. 2013). Remarkably, it was discovered that chemical spots in HgMn stars evolve slowly with time (Kochukhov et al. 2007)—something that has never been observed in magnetic Ap stars. Figure 9.3 shows an example of such an evolution, on a time scale of several years, for the mercury distribution in the brightest HgMn star α And. No conclusive theoretical explanation of the origin of these spots and reasons for their slow variation has been proposed. However, there are indications that a non-equilibrium, time-dependent atomic diffusion may play some role in these phenomena (Alecian et al. 2011).

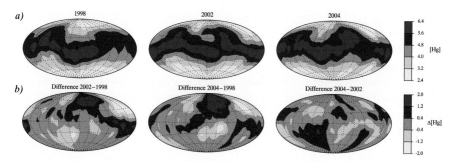

Fig. 9.3 Slow evolution of mercury overabundance spots on the surface of non-magnetic chemically peculiar star α And. (**a**) The Hammer-Aitoff projection of the Hg DI maps reconstructed at three different epochs spanning 7 years. The sidebar gives the local element abundance in logarithmic units relative to the solar concentration of mercury. (**b**) The corresponding pairwise difference maps. Adapted from Kochukhov et al. (2007)

Temperature Mapping of Active Late-Type Stars

In parallel with the DI studies of chemically peculiar stars significant efforts were devoted to reconstruction of temperature inhomogeneities on the surfaces of cool active stars. In these objects the surface activity is invariably associated with an enhanced dynamo action amplified by a rapid stellar rotation. This makes these stars much more active and consequently more spotted than the Sun. Several classes of late-type active stars exhibit these characteristics: single rapidly rotating pre-main sequence (including classical and weak-line T Tauri) and young main sequence stars, members of close binary systems spun up by tidal interaction (RS CVn stars), various classes of rapidly rotating giants (e.g. FK Com-type stars). The presence of large spots on the surfaces of these stars and significant rotational broadening of their spectral line profiles facilitate application of DI. However, unlike static chemical maps of Ap stars, temperature distributions of active cool stars evolve with time. This imposes a significant limitation of the observational data: a complete rotational phase coverage has to be achieved within ≤ 10 stellar rotations, which typically corresponds to a single observing run.

Temperature maps have been published for ∼ 100 cool stars. A catalogue and detailed review of these studies can be found in Strassmeier (2009, 2011). A few active stars were targeted repeatedly by DI (Vogt et al. 1999; Kovári et al. 2004; Korhonen et al. 2007; Hackman et al. 2012) in an effort to reveal activity cycles and compare them with predictions of dynamo theories. Despite collecting a valuable data base of temperature maps, these investigations achieved a rather limited success in establishing connections with dynamo theory and with well-known cyclic activity behaviour of the Sun. This is either because the most active cool stars usually chosen for DI do not exhibit well-defined activity cycles or because a long-term behaviour was sampled by DI studies with too few maps to reach definite conclusions about the nature of temporal evolution of the surface structure.

Methodologically, the temperature mapping of cool stars is commonly performed using a small number of diagnostic lines or even a single spectral feature. These spectral lines are modelled by assigning individual temperature values to each stellar surface element and using model atmospheres from standard grids with T_{eff} corresponding to these local temperatures to calculate local continuum and line intensities. Reliability of temperature distributions, especially the spot to photosphere temperature contrast ΔT_s, can be improved by considering atomic lines with different temperature sensitivity or by incorporating molecular bands in the temperature inversion (Rice et al. 2011). Summarising studies with the most careful determination of ΔT_s, Berdyugina (2005) found that there exists a systematic trend of the spot contrast with the stellar T_{eff}. The maximum $\Delta T_s \approx 2000 \, \text{K}$ is found for early G-type stars while the minimum contrast of only a few hundred K is seen in active M dwarfs.

Reconstruction of the stellar surface map in terms of brightness distribution or fractional spot coverage instead of local temperature is an alternative method of performing cool star DI (Collier Cameron 1998). These studies often completely neglect variation of the local line intensities with temperature, attributing all disk-integrated line profile variability to changes of the local continuum brightness. In other cases spectra of slowly rotating cool and hot template stars are employed to approximate the local spot and photospheric profiles of the DI target (Unruh et al. 1995). Due to this simplified spectrum synthesis approach results of the brightness DI studies cannot be directly interpreted in terms of physical parameters, such as spot temperatures. This type of DI also requires some external information (e.g. setting the spot to photosphere brightness contrast or choosing appropriate template stars) that can only be provided by more physically detailed studies. On the other hand, this method is well adapted to using high S/N ratio mean line profiles constructed by combining information from thousands of individual metal lines. This enables a fast and very precise (but not necessarily accurate) reconstruction of the stellar surface map, which is particularly suitable for detection of subtle secondary effects such as differential rotation (Donati et al. 2003; Barnes et al. 2005).

Mapping of Stellar Non-radial Pulsations

Stellar non-radial pulsations (NRP) represent another well-known cause of line profile variability. Pulsational perturbation on the stellar surface produces an alternating pattern of zones receding and approaching to the observer. The resulting velocity shifts are superimposed on the rotational Doppler shifts, producing a characteristic periodic variability of the disk-integrated line profiles (Vogt and Penrod 1983a).

Typically, the stellar NRP pattern is parameterised with the spherical harmonic functions and is fully described by specifying the ℓ and m numbers of each pulsation mode, and a ratio of the vertical to horizontal pulsation amplitude. However, in some particularly interesting cases the NRP geometry is significantly and non-trivially distorted by a rapid stellar rotation or a strong magnetic field (Lee and Saio

1990; Saio and Gautschy 2004). In such cases a mono-periodic pulsation cannot be described with a single combination of ℓ and m values. Instead, one can attempt to reconstruct the surface pulsation pattern by solving an ill-posed DI problem. For example, Berdyugina et al. (2003) mimicked the line profile variation due to NRP with temperature spots. They performed a pseudo-temperature DI for the B-type pulsating star ω^1 Sco with a dominant sectoral mode ($\ell = m$). In this particular case pulsations can be described as a fixed surface pattern that rotates with a period determined by the stellar rotation period and the m value (which has to be guessed) of the sectoral pulsation mode.

In another implementation of the NRP DI problem, Kochukhov (2004a) developed a more physically realistic description of an arbitrary pulsational velocity perturbation in terms of two surface maps, effectively representing the surface distribution of the pulsational amplitude and phase. This study demonstrated that it is possible to reconstruct both maps with the help of DI methodology, provided that the stellar rotational and pulsational periods are known. This DI technique was applied to the rapidly oscillating Ap star HR 3831 (Kochukhov 2004b), yielding the first DI NRP velocity map and providing a unique insight into the interplay between p-mode pulsations and a global magnetic field.

9.3 Zeeman Doppler Imaging with Polarisation Spectra

9.3.1 Detection and Diagnostic of Stellar Magnetic Field

The presence of a magnetic field leads to splitting of the atomic energy levels due to the Zeeman effect. Consequently, individual spectral lines corresponding to the transitions between Zeeman-split levels separate into groups of so-called π and σ components. The magnitude of this separation depends on the magnetic field strength, magnetic sensitivity of a given spectral line (characterised by the mean Landé factor), and the central wavelength of this line. Then, a magnetic field at the stellar surface can be detected by two basic effects: the splitting of magnetically sensitive lines and the presence of polarisation in Zeeman components.

Typical magnetic fields of non-degenerate stars produce Zeeman splitting which is much smaller than the intrinsic line width. Only very strong ($\geq$ 1–2 kG) magnetic fields of slowly rotating Ap stars and active M dwarfs can be diagnosed by the Zeeman splitting or broadening of spectral lines in high-dispersion optical spectra (Mathys et al. 1997; Reiners and Basri 2007). At the same time, magnetic field is normally the only cause of polarisation in spectral lines. This means that the mere presence of a systematic line polarisation signal is a signature of magnetic field.

The full state of polarisation of stellar radiation is characterised by the four Stokes parameters: Stokes I (total intensity), V (circular polarisation), and QU (linear polarisation). The line profile shape in Stokes I is primarily sensitive to the field modulus. The amplitude of the corresponding circular polarisation signal is given by the line of sight projection of the magnetic field vector. The magnitude of the transverse field component and its orientation in the plane of the sky determines the Stokes QU parameters. Thus, all four Stokes parameters are in principle needed for a complete diagnostic of stellar magnetic field. However, the Zeeman effect produces circular polarisation that is up to 10 times stronger than linear polarisation. In addition, the local Zeeman circular polarisation profile has a simpler morphology (S-shape, double-lobe signature) compared to the linear polarisation profiles (M or W-shape, three-lobe signature).[1] For these reasons, the vast majority of stellar magnetism studies rely exclusively on the Stokes V observations both for the field detection and modelling.

Even constrained to the Stokes V parameter, stellar spectropolarimetry is a challenging task owing to low amplitudes of the typical polarisation signals. For example, the global kG-strength magnetic fields of Ap stars produce circular polarisation signatures with amplitudes of a few % of the unpolarised continuum. These signatures can be observed in individual spectral lines (e.g. Silvester et al. 2012) provided that the spectra have a S/N ratio of 300–500. Much weaker magnetic fields of active cool stars yield disk-integrated signals at the level of 10^{-3} to 10^{-4} in Stokes V. A detection of 1–10 G global fields of moderately active or inactive stars, such as the Sun, requires polarimetric precision of 10^{-5}–10^{-6}. Obviously, such signals are impossible to detect in individual spectral lines. Instead, efficient multi-line polarisation methods were developed to average polarisation signatures over all suitable spectral lines. These procedures reduce the photon noise to the level required for the detection of extremely weak polarisation signals. The most commonly used multi-line method of least-squares deconvolution (LSD, Donati et al. 1997; Kochukhov et al. 2010) incorporates de-blending and recovers an average profile shape that in the context of DI can often be treated as a single spectral line with mean parameters.

A fundamental difference between magnetic field signatures in the intensity and polarisation is that the Zeeman splitting in the Stokes I spectra depends on the field modulus while the Stokes QUV profiles are highly sensitive to the field orientation. On the one hand, this means that polarisation observations are providing rich information about field geometry. On the other hand, any analysis of complex magnetic topologies inevitably suffers from a cancellation of polarisation signals in the disk-integrated spectra due to addition of polarisation signatures with opposite signs corresponding to the regions of opposite field polarity.

---

[1] We refer to the textbook by Landi Degl'Innocenti and Landolfi (2004) for a comprehensive review of spectral line polarisation theory.

9.3.2 Mapping of Stellar Magnetic Field Topologies

Magnetic field is a vector quantity. Therefore, to fully describe the stellar surface magnetic field topology one has to specify three scalar two-dimensional maps of the radial, meridional, and azimuthal magnetic field components. Simultaneous reconstruction of these distributions from spectropolarimetric observations is a formidable task, generally requiring data in all four Stokes parameters. Such observations are currently available for a small sample of strongly magnetic Ap stars (Wade et al. 2000; Silvester et al. 2012; Rusomarov et al. 2013) and for only one bright active RS CVn star (Rosén et al. 2013).

Figure 9.4 gives an example of the Stokes *IQUV* profile variation for a star with a strong dipolar magnetic field. These theoretical profiles demonstrate that magnetic field influences the line shapes in intensity as well as gives rise to a stronger circular and weaker linear polarisation signatures. Piskunov and Kochukhov (2002) and Kochukhov and Piskunov (2002) showed that based on an observational data set comprising 10–20 phase of high-quality Stokes *IQUV* spectra it is possible to recover a vector magnetic field map using the same basic mathematical formulation of the DI problem as described in Sect. 9.2.3 and applying Tikhonov regularisation to each of the three magnetic field component maps individually. In their magnetic inversion method Piskunov and Kochukhov (2002) modelled the Stokes parameter profiles of individual spectral lines with realistic numerical polarised radiative transfer calculations and implemented a self-consistent and simultaneous magnetic

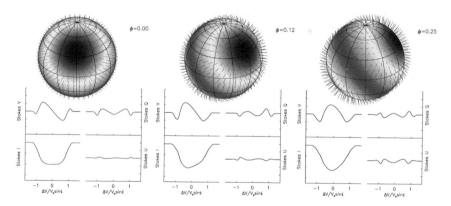

Fig. 9.4 Stokes profile variation of a star with an oblique dipolar magnetic field. The spherical plots show the surface magnetic field distribution for three different rotational phases. The underlying colour image represents the field modulus. The vector map shows the field orientation with different colours highlighting areas of the inward and outward directed magnetic field. The Stokes *IQUV* profiles are illustrated below. The scale of the polarisation profiles is the same for Stokes *V* and *QU* but is expanded by a factor of 1.4 relative to Stokes *I*. These calculations are for 4 kG dipolar field inclined by $\beta = 50°$ with respect to the stellar rotational axis. The latter is inclined by $i = 60°$ with respect to the line of sight. The projected stellar rotational velocity is 30 km s^{-1}

and chemical spot inversions. According to the numerical tests by Kochukhov and Piskunov (2002), this *general* ZDI methodology is successful in recovering both global and structured magnetic field topologies. This ZDI method is now routinely used to study the field topologies and chemical spot distributions in magnetic Ap stars (see Sect. 9.3.3).

However, due to lack of linear polarisation observations for cool magnetic stars, a different *restricted* and less sophisticated ZDI technique is usually applied. Brown et al. (1991) and Donati et al. (1997) developed and tested a MEM-based inversion procedure relying on the Stokes V time series, which still aims at recovering all three components of the stellar magnetic field distribution. Figure 9.5 gives an idea of how one can constrain the field orientation with the help of rotational modulation of the Stokes V signatures of magnetic spots. In this highly simplified example of isolated circular magnetic spots, the radial field spot produces a Stokes V signature moving across the stellar line profile from blue to red and showing a maximum amplitude when the spot is at the disk centre (Fig. 9.5a). The meridional field of the same strength produces a much weaker Stokes V signal that has a minimum at the disk centre (Fig. 9.5b). On the other hand, the azimuthal field exhibits a noticeably different Stokes V signature, which changes sign at the disk centre (Fig. 9.5c). Based on this behaviour, one expects that the azimuthal field can be readily distinguished from the radial and meridional fields. But the latter two magnetic components are difficult to disentangle from each other over at least some part of the stellar surface. Reconstruction of the meridional field is going to be the least reliable. Detailed ZDI numerical tests by Brown et al. (1991), Kochukhov and Piskunov (2002) and Rosén and Kochukhov (2012) confirmed this assessment.

Returning to the question of regularisation in ZDI, Brown et al. (1991) found that the restricted Stokes V inversions produce reasonable results for isolated radial field spots but fail for global magnetic field distributions such as a dipolar field. This problem was attributed to the fundamental assumption of the MEM regularisation (Piskunov and Kochukhov 2002): it requires setting a "default" value that cannot be easily and uniquely defined for a global magnetic field geometry. In contrast, the Tikhonov regularisation performs much better for global magnetic geometries. More recent ZDI studies switched to using a spherical harmonic representation of the stellar magnetic field (Donati et al. 2006; Kochukhov et al. 2014). In this case a ZDI inversion code recovers spherical harmonic coefficients corresponding to the poloidal and toroidal magnetic field components rather than the local magnetic field values. This modification allows one to model both global and highly structured magnetic topologies with an added benefit of being able to characterise the field in detail (e.g. assess poloidal vs. toroidal or axisymmetric vs. non-axisymmetric field). In this case the regularisation is accomplished by limiting the maximum angular degree ℓ of the harmonic expansion and by adding to chi-square an additional

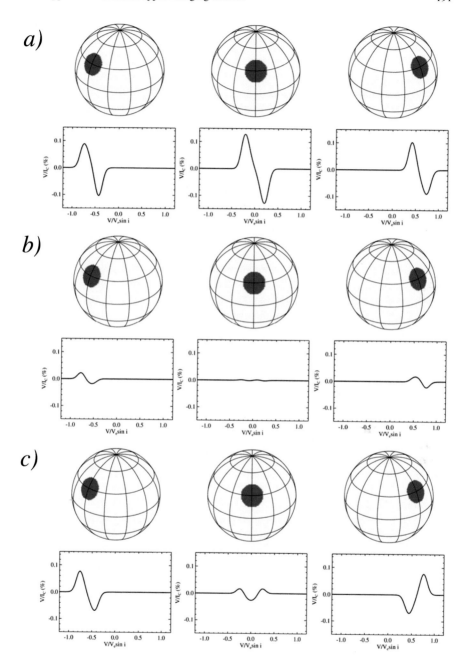

Fig. 9.5 Signatures of magnetic spots in the circular polarisation line profiles of a rotating star. Different rows show the Stokes V profiles corresponding to the radial (**a**), meridional (**b**), and azimuthal (**c**) magnetic spots with the same field strength. In each case the star is shown at three rotational phases, separated by 0.125 of the rotational period

penalty function which discourages the inversion code to introduce high-order harmonic modes not justified by the observational data.[2]

The restricted Stokes V inversions are usually applied to the LSD circular polarisation profiles of cool active stars using a simplified polarised line formation models. Typically, a Gaussian local line profile together with the weak-field approximation (Marsden et al. 2011) or the Milne-Eddington analytical solution of the polarised radiative transfer equation (Morin et al. 2008) is used. Except for the recent series of studies of global magnetic fields of classical T Tauri stars (Donati et al. 2010) and a couple of exploratory analyses of other types of active stars (Carroll et al. 2012; Kochukhov et al. 2013), these ZDI studies completely ignore the effect of cool spots on the circular polarisation profiles. Consequently, the field strength in the vicinity of dark features is at least significantly underestimated. In the worst-case scenario when all magnetic field is concentrated within cool spots magnetic field distributions inferred by this inversion methodology are largely spurious (Rosén and Kochukhov 2012).

A new self-consistent ZDI inversion method based on detailed polarised radiative transfer modelling of the LSD profiles in two or all four Stokes parameters was introduced by Kochukhov et al. (2014). This technique is computationally expensive but overcomes most of the shortcomings of the restricted ZDI based on LSD profiles and can be applied to both early- and late-type magnetic stars.

9.3.3 ZDI Applications

Magnetic Mapping of B and A Stars

Historically information about magnetic field topologies of Ap stars was obtained by fitting the phase curves of integral magnetic observables (longitudinal magnetic field, mean field modulus, etc.) with low-order multipolar geometries (Landstreet and Mathys 2000; Bagnulo et al. 2002). These studies established that the majority of early-type magnetic stars possess nearly dipolar fields, with occasional deviations from axisymmetry or a quadrupolar contribution (Kochukhov 2006; Bailey et al. 2012). Initial applications of ZDI to high-resolution Stokes I and V observations of Ap stars validated this picture (Kochukhov et al. 2002, 2014; Lüftinger et al. 2010). Even using detailed circular polarisation profile modelling of individual spectral lines in the framework of the general ZDI methodology discussed above little deviation from the basic dipolar field topology was found for A-type magnetic stars.

---

[2]Somewhat confusingly, such ZDI with the spherical harmonic regularisation is still commonly called "maximum entropy" inversion although the employed regularisation is quite different from the original MEM-based ZDI scheme described by Brown et al. (1991).

A qualitatively different picture emerged when ZDI was applied, for the first time, to the full Stokes vector observations of Ap stars obtained with the MuSiCoS spectropolarimeter (Wade et al. 2000). In particular, Kochukhov et al. (2004) and Kochukhov and Wade (2010) showed that satisfactory fits to the Stokes QU (linear polarisation) observations of Ap stars 53 Cam and α^2 CVn require a considerably more complex field topologies than is apparent from the Stokes IV profiles of these stars. Figure 9.6 shows a comparison of the ZDI results by Kochukhov and Wade (2010) for α^2 CVn obtained by considering all four Stokes parameter spectra (Fig. 9.6a) and excluding the linear polarisation data from inversion (Fig. 9.6b). Evidently, the full Stokes vector ZDI is able to recover the small-scale magnetic field structures—essentially comprising several horizontal field spots—that remain undetected by the inversions limited to Stokes IV data.

Subsequent investigation of α^2 CVn using high-quality Stokes $IQUV$ observations secured with a new generation of spectropolarimeters (ESPaDOnS and Narval, see Silvester et al. 2014) confirmed the ZDI results of Kochukhov and Wade (2010)

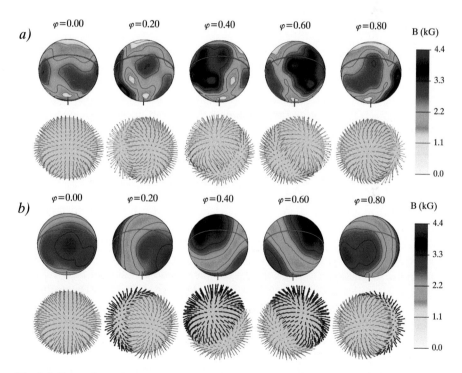

Fig. 9.6 Comparison of the ZDI reconstruction of the surface magnetic field topology of the Ap star α^2 CVn from observations in all four Stokes parameters (**a**) and from a data set limited to Stokes I and V (**b**). In both cases the rows show spherical plots of the magnetic field modulus (*upper row*) and field orientation (*lower row*). The star is shown at five rotational phases and at the inclination angle of $i = 120°$. The contours are plotted with a 0.5 kG step in the field strength map. It is evident that the magnetic inversion limited to the circular polarisation spectra is unable to recover the small-scale magnetic features. Adapted from Kochukhov and Wade (2010)

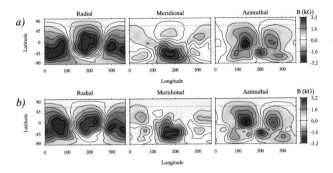

Fig. 9.7 Long-term stability of the global and small-scale magnetic field topology of the Ap star α^2 CVn. Rectangular plots show surface distributions of the radial, meridional, and azimuthal field components. The two rows correspond to (**a**) magnetic field map derived by Kochukhov and Wade (2010) from the MuSiCoS observations collected in 1997–1999 and (**b**) an equivalent magnetic field distribution obtained by Silvester et al. (2012) from the ESPaDOnS and Narval four Stokes parameter data acquired in 2006–2007. In both cases magnetic inversions were carried out using all four Stokes parameters. Adapted from Silvester et al. (2012)

and demonstrated that no significant evolution of either the global or small-scale magnetic field of this star has occurred during ~ 10 years that have passed between acquisition of the two polarisation data sets (see Fig. 9.7).

Thus, the four Stokes parameter ZDI studies of intermediate-mass Ap stars 53 Cam and α^2 CVn indicated that, although the overall magnetic field structure defined by the radial field component is dipolar-like, there are also smaller scale magnetic features on the stellar surface. Since very few Ap stars were studied with four Stokes parameter ZDI, it is difficult to ascertain how this picture depends on the stellar mass and age. To this end, recent results seem to hint on the mass (or age) dependence of the level of complexity of Ap star magnetic fields. Rusomarov et al. (2015) reported no evidence of deviations from a dipolar field topology based on the ZDI analysis of very high quality HARPSpol Stokes $IQUV$ spectra of the low-mass (and old) Ap star HD 24712. On the other hand, a couple of magnetic early B stars (hence massive and young) were found to possess remarkable non-dipolar magnetic geometries from a complex phase variation of their circular polarisation profiles (Donati et al. 2006; Kochukhov et al. 2011). ZDI results for one of these extraordinary stars, HD 37776, are presented in Fig. 9.8. In this case the field structure does not resemble a dipole or quadrupole even remotely. So far this extreme degree of the field complexity has only been seen in the most massive early-type magnetic stars.

Several ongoing four Stokes parameter ZDI studies will increase the sample of Ap stars with detailed information on the surface magnetic field structure, helping to clarify the emerging trend of the field complexity with stellar parameters.

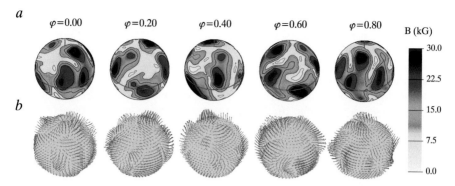

Fig. 9.8 Unusually complex magnetic field structure of the B-type star HD 37776 reconstructed with ZDI. The spherical maps show surface distributions of (**a**) the field modulus and (**b**) field orientation. This magnetic field geometry clearly deviates from a simple axisymmetric dipole or dipole+quadrupole topology. From Kochukhov et al. (2011)

Magnetic Mapping of Late-Type Active Stars

First ZDI maps of the magnetic field topologies of cool active stars were published by Donati and Collier Cameron (1997), Donati (1999) and Donati et al. (1999, 2003). These authors studied a couple of rapidly rotating single dwarf stars (AB Dor, LQ Hya) and the primary giant component of the RS CVn binary HR 1099. For all three objects inversions were performed for several epochs, giving an idea of the field evolution. These ZDI studies employed LSD profiles for the restricted ZDI reconstruction (see above) of the brightness map from Stokes I and the magnetic field distribution from Stokes V data.

Magnetic geometries of cool active stars turned out to be qualitatively different from that of the Sun. Rather than featuring a system of bipolar regions with mostly radial field orientation, as was expected by some "active Sun" models (Schrijver and Title 2001), the first ZDI targets showed significant amounts of large-scale horizontal magnetic fields. This field was found to be arranged in azimuthal bands which evolved with time (Donati 1999). Another surprising result was the lack of correlation between the dark photospheric regions and magnetic features, suggesting that the strong magnetic fields inside star spots are not resolved or that the restricted ZDI technique is not sensitive to such fields (Donati and Collier Cameron 1997). In general, the local magnetic field strengths inferred by ZDI studies occasionally reach 1 kG in the strongest magnetic concentrations but, more typically, amount to only a few hundred G. However, extrapolating from the physics of cool spot formation on the solar surface, multi-kG fields should be ubiquitous in active stars with large star spots.

Subsequently ZDI analyses were applied to many other classes of late-type active stars, ranging from F stars to M dwarfs. A comprehensive review of these investigations can be found in Donati and Landstreet (2009). One of the most impressive achievements was an extension of the magnetic mapping to G-type stars with the overall activity levels and rotation rates comparable to the Sun (Petit et al. 2008). This work demonstrated that magnetic inversions can constrain the global field topologies even for stars with very small $v_e \sin i$. As discussed above (Sect. 9.2.2), in that case information is primarily extracted from the rotational modulation of the Stokes V signal rather than from Doppler shifts. For a sample of four solar analogues Petit et al. (2008) found that the balance between poloidal and toroidal contributions to the global magnetic field geometry depends on the stellar rotation rate. Stars rotating faster than 12 days show predominantly toroidal field and stars with a slower rotation exhibit poloidal field topology, reminiscent of the global structure of the solar magnetic field.

A few active stars were systematically followed by ZDI over several years (Fares et al. 2009; Morgenthaler et al. 2012; Kochukhov et al. 2013). In most cases a significant change of the global field topology was detected. For example, Fig. 9.9 illustrates the magnetic field reversal for the RS CVn star II Peg studied by Kochukhov et al. (2013). It remains to be seen how these direct observations of the magnetic field cycles relate to the behaviour of indirect magnetic proxies such as the X-ray and Ca H&K emission measures. So far no clear link between the cycles in direct magnetic observations and proxy indicators was found.

A new type of stellar magnetic field topologies was identified in active M dwarfs stars with the help of ZDI (Morin et al. 2008; Donati et al. 2008). It turns out that the convective dynamo mechanism operating in mid- and late-M dwarfs produces fairly strong (~ 500 G), large-scale magnetic fields. For the majority of these stars the field topology is dipolar and aligned with the stellar rotational axis. On the other hand, early M dwarfs tend to exhibit more complex, weaker and non-axisymmetric fields. At the same time, it also became clear that the Stokes V ZDI of M dwarfs misses up to 95 % of the magnetic flux (Reiners and Basri 2009) because the field modulus measured from the Stokes I spectra of the same objects indicates 2–4 kG fields (Johns-Krull et al. 1999; Reiners and Basri 2007), which is much larger than the global field strength inferred by polarimetry. Presumably, these strong fields have a complex structure and therefore cancel out in the disk-integrated polarisation signal. Quantitative field topology models simultaneously reproducing both the Stokes I and V observations of M dwarfs are yet to be developed.

Over the past few years significant efforts were made to test key assumptions of the restricted ZDI inversions and to introduce more realistic polarisation modelling methodologies in the ZDI of cool stars. Limitations of the traditional single-line interpretation of the Stokes V LSD profiles were explored by Kochukhov et al. (2010). Based on the experience gained from the general ZDI of Ap stars, Kochukhov et al. (2013) performed ZDI inversions for the RS CVn star II Peg using detailed self-consistent polarised radiative transfer calculations. According to the numerical tests by Rosén and Kochukhov (2012), this should have been sufficient for identifying strong unipolar magnetic fields inside cool spots. However, such fields were not detected suggesting that the dark spots recovered by DI are not monolithic but are composed of numerous bipolar groups whose polarisation signals cancel out in the disk-integrated Stokes V spectra.

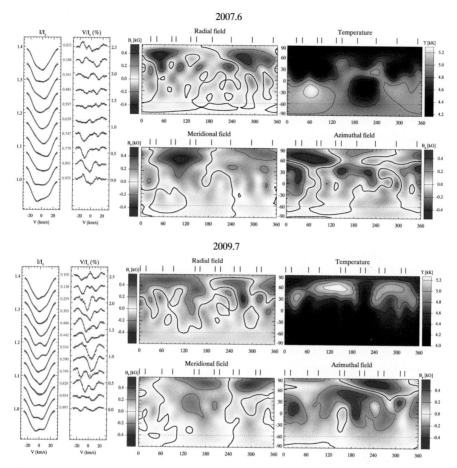

Fig. 9.9 Self-consistent ZDI reconstruction of the vector magnetic field and temperature distribution for the RS CVn star II Peg at two different epochs. For each epoch the two columns on the left compare the observed (*symbols*) and theoretical (*lines*) LSD Stokes *I* and *V* profiles. The four rectangular maps illustrate distributions of the radial, meridional, and azimuthal field components, and temperature. Significant evolution of the surface structure over a time span of 2 years is evident. From Kochukhov et al. (2013)

In another development, Kochukhov et al. (2011) and Rosén et al. (2013) obtained the first spectropolarimetric observations of cool active stars in all four Stokes parameters. It was demonstrated that the full Stokes vector Zeeman spectropolarimetry is feasible using LSD, at least for a handful of brightest objects. A ZDI investigation based on these data was published by Rosén et al. (2015). Their results indicate that, similar to the situation with ZDI of Ap stars, simultaneous modelling of the Stokes *IQUV* spectra reveals considerably more complex magnetic fields than suggested by the traditional Stokes *V* inversions.

9.4 Conclusions

Doppler and Zeeman Doppler imaging have proven themselves as a powerful remote sensing methods of obtaining detailed maps of spot distributions and magnetic field topologies for unresolved stellar surfaces. DI and ZDI have been applied to a large number of stars, leading to several important breakthroughs in our understanding of the stellar magnetism and the processes of surface structure formation. As a conclusion of this review we take a look into the future, summarising key development directions of Doppler inversion studies of early- and late-type stars.

- **DI of chemical spots** Doppler inversions of chemical abundance inhomogeneities in chemically peculiar stars recover horizontal distributions under the assumption of no significant vertical abundance variation. However, Ap stars are known to exhibit vertical stratification of chemical elements (Ryabchikova et al. 2002) and the apparent lateral inhomogeneities may well be a consequence of the variation of vertical stratification over the stellar surface (Alecian and Stift 2010). Therefore, chemical abundance DI should eventually incorporate the vertical dimension in the inversion process, ultimately providing 3-D chemical spot maps.
- **DI of temperature spots** This application of DI would benefit from an increase of reliability of reconstruction of the physical properties of star spot interiors. This can be accomplished by a systematic incorporation of the molecular indicators in the temperature DI modelling and combining the optical and near-infrared spectroscopic diagnostics. At some stage, the question of the vertical temperature and pressure structure of magnetised star spots needs to be addressed and dedicated models of spot atmospheres be developed and incorporated in DI.
- **ZDI of early-type magnetic stars** The methodology of self-consistent four Stokes parameter inversions using polarisation profiles of individual lines is well-established and thoroughly tested. However, this technique requires Stokes $IQUV$ observational data of superb quality that is unavailable for all but brightest stars. It is of interest to pursue development of multi-line four Stokes parameter ZDI methodology, using the S/N gain advantages of LSD but without compromising on the detailed polarised radiative transfer calculations. First steps in this direction were taken in the Stokes IV inversions described by Kochukhov et al. (2014), but the method is yet to be tested for a full Stokes vector data set.
- **ZDI of cool active stars** Despite numerous applications of magnetic mapping to different classes of active stars, there remain fundamental questions about reliability of the ZDI inversions of complex field topologies using only Stokes IV observations. Furthermore, the consequences of numerous simplifications of the restricted ZDI (analytical line profiles, lack of self-consistency between spot and magnetic field modelling, etc.) remain poorly explored. Occasional comparisons of the ZDI reconstructions by independent inversion codes reveal uncomfortably large discrepancies (see Skelly et al. 2010; Carroll et al. 2012). Methodological improvements of the cool star ZDI are essential for understanding the absence

of correlation between temperature and magnetic maps and the apparent lack of strong magnetic fields that should be associated with cool star spots.

References

Adelman, S. J., Gulliver, A. F., Kochukhov, O. P., & Ryabchikova, T. A. (2002, August). The variability of the Hg II λ3984 line of the mercury-manganese star α andromedae. *Astrophysical Journal, 575*, 449–460.

Alecian, G., & Stift, M. J. (2010, June). Bi-dimensional element stratifications computed for magnetic Ap star atmospheres. *Astronomy and Astrophysics, 516*, A53

Alecian, G., Stift, M. J., & Dorfi, E. A. (2011, December). Time-dependent diffusion in stellar atmospheres. *Monthly Notices of the Royal Astronomical Society, 418*, 986–997.

Alecian, G., & Vauclair, S. (1981, August). Element stratification in the atmospheres of main sequence stars - The silicon accumulation. *Astronomy and Astrophysics, 101*, 16–25.

Aurière, M., Wade, G. A., Lignières, F., Hui-Bon-Hoa, A., Landstreet, J. D., Iliev, I. K., et al. (2010, November). No detection of large-scale magnetic fields at the surfaces of Am and HgMn stars. *Astronomy and Astrophysics, 523*, A40.

Bagnulo, S., Landi Degl'Innocenti, M., Landolfi, M., & Mathys, G. (2002, November). A statistical analysis of the magnetic structure of CP stars. *Astronomy and Astrophysics, 394*, 1023–1037.

Bagnulo, S., Landstreet, J. D., Fossati, L., & Kochukhov, O. (2012, February). Magnetic field measurements and their uncertainties: the FORS1 legacy. *Astronomy and Astrophysics, 538*, A129.

Bailey, J. D., Grunhut, J., Shultz, M., Wade, G., Landstreet, J. D., Bohlender, D., et al. (2012, June). An analysis of the rapidly rotating Bp star HD 133880. *Monthly Notices of the Royal Astronomical Society, 423*, 328–343.

Barnes, J. R., Collier Cameron, A., Donati, J.-F., James, D. J., Marsden, S. C., & Petit, P. (2005, February). The dependence of differential rotation on temperature and rotation. *Monthly Notices of the Royal Astronomical Society, 357*, L1-L5.

Berdyugina, S. V. (2005, December). Starspots: A key to the stellar dynamo. *Living Reviews in Solar Physics, 2*, 8.

Berdyugina, S. V., Telting, J. H., Korhonen, H., & Schrijvers, C. (2003, July). Surface imaging of stellar non-radial pulsations. II. The β Cephei star ω¹ Sco. *Astronomy and Astrophysics, 406*, 281–285.

Briquet, M., Korhonen, H., González, J. F., Hubrig, S., & Hackman, T. (2010, February). Dynamical evolution of titanium, strontium, and yttrium spots on the surface of the HgMn star HD 11753. *Astronomy and Astrophysics, 511*, A71.

Brown, S. F., Donati, J.-F., Rees, D. E., & Semel, M. (1991, October). Zeeman-Doppler imaging of solar-type and AP stars. IV - Maximum entropy reconstruction of 2D magnetic topologies. *Astronomy and Astrophysics*, 463–474.

Carroll, T. A., Strassmeier, K. G., Rice, J. B., & Künstler, A. (2012, December). The magnetic field topology of the weak-lined T Tauri star V410 Tauri. New strategies for Zeeman-Doppler imaging. *Astronomy and Astrophysics, 548*, A95.

Collier Cameron, A. (1998). Stellar tomography. *Astrophysics and Space Science, 261*, 71–80.

Donati, J., & Collier Cameron, A. (1997, October). Differential rotation and magnetic polarity patterns on AB Doradus. *Monthly Notices of the Royal Astronomical Society, 291*, 1–19.

Donati, J.-F. (1999, January). Magnetic cycles of HR 1099 and LQ Hydrae. *Monthly Notices of the Royal Astronomical Society, 302*, 457–481.

Donati, J.-F., Cameron, A. C., Semel, M., Hussain, G. A. J., Petit, P., Carter, B. D., et al. (2003, November). Dynamo processes and activity cycles of the active stars AB Doradus, LQ Hydrae and HR 1099. *Monthly Notices of the Royal Astronomical Society, 345*, 1145–1186.

Donati, J.-F., Collier Cameron, A., Hussain, G. A. J., & Semel, M. (1999, January). Magnetic topology and prominence patterns on AB Doradus. *Monthly Notices of the Royal Astronomical Society, 302*, 437–456.

Donati, J.-F., Collier Cameron, A., & Petit, P. (2003, November). Temporal fluctuations in the differential rotation of cool active stars. *Monthly Notices of the Royal Astronomical Society, 345*, 1187–1199.

Donati, J.-F., Howarth, I. D., Jardine, M. M., Petit, P., Catala, C., Landstreet, J. D., et al. (2006, August). The surprising magnetic topology of τ Sco: fossil remnant or dynamo output? *Monthly Notices of the Royal Astronomical Society, 370*, 629–644.

Donati, J.-F., & Landstreet, J. D. (2009, September) Magnetic fields of nondegenerate stars. *The Annual Review of Astronomy and Astrophysics, 47*, 333–370.

Donati, J.-F., Morin, J., Petit, P., Delfosse, X., Forveille, T., Aurière, M., et al. (2008, September) Large-scale magnetic topologies of early M dwarfs. *Monthly Notices of the Royal Astronomical Society, 390*, 545–560.

Donati, J.-F., Semel, M., Carter, B. D., Rees, D. E., & Collier Cameron, A. (1997, November). Spectropolarimetric observations of active stars. *Monthly Notices of the Royal Astronomical Society, 291*, 658–682.

Donati, J.-F., Skelly, M. B., Bouvier, J., Gregory, S. G., Grankin, K. N., Jardine, M. M., et al. (2010, December). Magnetospheric accretion and spin-down of the prototypical classical T Tauri star AA Tau. *Monthly Notices of the Royal Astronomical Society, 409*, 1347–1361.

Fares, R., Donati, J.-F., Moutou, C., Bohlender, D., Catala, C., Deleuil, M., et al. (2009, September) Magnetic cycles of the planet-hosting star τ Bootis - II. A second magnetic polarity reversal. *Monthly Notices of the Royal Astronomical Society, 398*, 1383–1391.

Folsom, C. P., Kochukhov, O., Wade, G. A., Silvester, J., & Bagnulo, S. (2010, October). Magnetic field, chemical composition and line profile variability of the peculiar eclipsing binary star AR Aur. *Monthly Notices of the Royal Astronomical Society, 407*, 2383–2392.

Goncharskii, A. V., Ryabchikova, T. A., Stepanov, V. V., Khokhlova, V. L., & Yagola, A. G. (1983, February). Mapping of elements on the surfaces of Ap-stars - part two - distribution of eu Sr and Si on alpha-2-canum-venaticorum Ch-serpentis and Cu-virginis. *Soviet Astronomy, 27*, 49–53.

Goncharskii, A. V., Stepanov, V. V., Kokhlova, V. L., & Yagola, A. G. (1977). Reconstruction of local line profiles from those observed in an Ap spectrum. *Soviet Astronomy Letters, 3*, 147–149.

Hackman, T., Mantere, M. J., Lindborg, M., Ilyin, I., Kochukhov, O., Piskunov, N., et al. (2012, February). Doppler images of II Pegasi for 2004–2010. *Astronomy and Astrophysics, 538*, A126.

Hatzes, A. P. (1991, February). Doppler images of abundance features on Theta Aurigae. *Monthly Notices of the Royal Astronomical Society, 248*, 487–493.

Hatzes, A. P. (1997, June). Doppler imaging of the silicon distribution on CUVir: Evidence for a decentred magnetic dipole? *Monthly Notices of the Royal Astronomical Society, 288*, 153–160.

Hubrig, S., González, J. F., Savanov, I., Schöller, M., Ageorges, N., Cowley, C. R., et al. (2006, October). Inhomogeneous surface distribution of chemical elements in the eclipsing binary ARAur: A new challenge for our understanding of HgMn stars. *Monthly Notices of the Royal Astronomical Society, 371*, 1953–1958.

Johns-Krull, C. M., Valenti, J. A., & Koresko, C. (1999, May). Measuring the magnetic field on the classical T Tauri star BP Tauri. *Astrophysical Journal, 516*, 900–915.

Khokhlova, V. L., & Pavlova, V. M. (1984). Maps of the iron-group elements on the magnetic Ap-star alpha-2-canum-venaticorum. *Soviet Astronomy Letters, 10*, 158–163.

Khokhlova, V. L., Rice, J. B., & Wehlau, W. H. (1986, August). Distribution of chemical elements over the surface of the magnetic AP star Theta Aurigae. *Astrophysical Journal, 307*, 768–776.

Kochukhov, O. (2004a, August). Doppler imaging of stellar non-radial pulsations. I. Techniques and numerical experiments. *Astronomy and Astrophysics, 423*, 613–628.

Kochukhov, O. (2004b, November). Indirect imaging of nonradial pulsations in a rapidly oscillating Ap star. *Astrophysical Journal, 615*, L149–L152.

Kochukhov, O. (2006, July). Remarkable non-dipolar magnetic field of the Bp star HD 137509. *Astronomy and Astrophysics, 454*, 321–325.

Kochukhov, O., Adelman, S. J., Gulliver, A. F., & Piskunov, N. (2007, August). Weather in stellar atmosphere revealed by the dynamics of mercury clouds in α Andromedae. *Nature Physics, 3*, 526–529.

Kochukhov, O., Bagnulo, S., Wade, G. A., Sangalli, L., Piskunov, N., Landstreet, J. D., et al. (2004, February). Magnetic doppler imaging of 53 camelopardalis in all four stokes parameters. *Astronomy and Astrophysics, 414*, 613–632.

Kochukhov, O., Drake, N. A., Piskunov, N., & de la Reza, R. (2004, September) Multi-element abundance Doppler imaging of the rapidly oscillating Ap star HR 3831. *Astronomy and Astrophysics, 424*, 935–950.

Kochukhov, O., Lüftinger, T., Neiner, C., Alecian, E., & MiMeS Collaboration. (2014, May). Magnetic field topology of the unique chemically peculiar star CU Virginis. *Astronomy and Astrophysics, 565*, A83.

Kochukhov, O., Lundin, A., Romanyuk, I., & Kudryavtsev, D. (2011, January). The extraordinary complex magnetic field of the helium-strong star HD 37776. *Astrophysical Journal, 726*, 24.

Kochukhov, O., Makaganiuk, V., & Piskunov, N. (2010, December). Least-squares deconvolution of the stellar intensity and polarization spectra. *Astronomy and Astrophysics, 524*, A5.

Kochukhov, O., Makaganiuk, V., Piskunov, N., Jeffers, S. V., Johns-Krull, C. M., Keller, C. U., et al. (2013, June). Are there tangled magnetic fields on HgMn stars? *Astronomy and Astrophysics, 554*, A61.

Kochukhov, O., Makaganiuk, V., Piskunov, N., Snik, F., Jeffers, S. V., Johns-Krull, C. M., et al. (2011, May). First detection of linear polarization in the line profiles of active cool stars. *Astrophysical Journal, 732*, L19.

Kochukhov, O., Mantere, M. J., Hackman, T., & Ilyin, I. (2013, February). Magnetic field topology of the RS CVn star II Pegasi. *Astronomy and Astrophysics, 550*, A84.

Kochukhov, O., & Piskunov, N. (2002, June). Doppler imaging of stellar magnetic fields. II. Numerical experiments. *Astronomy and Astrophysics, 388*, 868–888.

Kochukhov, O., Piskunov, N., Ilyin, I., Ilyina, S., & Tuominen, I. (2002, July). Doppler Imaging of stellar magnetic fields. III. Abundance distribution and magnetic field geometry of α² CVn. *Astronomy and Astrophysics, 389*, 420–438.

Kochukhov, O., Piskunov, N., Sachkov, M., & Kudryavtsev, D. (2005, September) Inhomogeneous distribution of mercury on the surfaces of rapidly rotating HgMn stars. *Astronomy and Astrophysics, 439*, 1093–1098.

Kochukhov, O., & Wade, G. A. (2010, April). Magnetic Doppler imaging of α² Canum Venaticorum in all four Stokes parameters. Unveiling the hidden complexity of stellar magnetic fields. *Astronomy and Astrophysics, 513*, A13.

Korhonen, H., Berdyugina, S. V., Hackman, T., Ilyin, I. V., Strassmeier, K. G., & Tuominen, I. (2007, December). Study of FK Comae Berenices. V. Spot evolution and detection of surface differential rotation. *Astronomy and Astrophysics, 476*, 881–891.

Korhonen, H., González, J. F., Briquet, M., Flores Soriano, M., Hubrig, S., Savanov, I., et al. (2013, May). Chemical surface inhomogeneities in late B-type stars with Hg and Mn peculiarity. I. Spot evolution in HD 11753 on short and long time scales. *Astronomy and Astrophysics, 553*, A27.

Kovári, Z., Strassmeier, K. G., Granzer, T., Weber, M., Oláh, K., & Rice, J. B. (2004, April). Doppler imaging of stellar surface structure. XXII. Time-series mapping of the young rapid rotator LQ Hydrae. *Astronomy and Astrophysics, 417*, 1047–1054.

Landi Degl'Innocenti, E., & Landolfi, M. (2004). *Polarization in Spectral Lines* (Vol. 307). Dordrecht: Kluwer Academic Publishers.

Landstreet, J. D., & Mathys, G. (2000, July). Magnetic models of slowly rotating magnetic Ap stars: aligned magnetic and rotation axes. *Astronomy and Astrophysics, 359*, 213–226.

Lee, U., & Saio, H. (1990, February). Line profile variations caused by low-frequency nonradial pulsations of rapidly rotating stars. *Astrophysical Journal, 349*, 570–579.

Lüftinger, T., Kochukhov, O., Ryabchikova, T., Piskunov, N., Weiss, W. W., & Ilyin, I. (2010, January). Magnetic Doppler imaging of the roAp star HD 24712. *Astronomy and Astrophysics, 509*(26), A71.

Makaganiuk, V., Kochukhov, O., Piskunov, N., Jeffers, S. V., Johns-Krull, C. M., Keller, C. U., et al. (2011, May). Chemical spots in the absence of magnetic field in the binary HgMn star 66 Eridani. *Astronomy and Astrophysics, 529*, A160.

Makaganiuk, V., Kochukhov, O., Piskunov, N., Jeffers, S. V., Johns-Krull, C. M., Keller, C. U., et al. (2012, March). Magnetism, chemical spots, and stratification in the HgMn star φ Phoenicis. *Astronomy and Astrophysics, 539*, A142.

Marsden, S. C., Jardine, M. M., Ramírez Vélez, J. C., Alecian, E., Brown, C. J., Carter, B. D., et al. (2011, May). Magnetic fields and differential rotation on the pre-main sequence - I. The early-G star HD 141943 - brightness and magnetic topologies. *Monthly Notices of the Royal Astronomical Society, 413*, 1922–1938.

Mathys, G., Hubrig, S., Landstreet, J. D., Lanz, T., & Manfroid, J. (1997, June). The mean magnetic field modulus of AP stars. *Astronomy and AstrophysicsS, 123*, 353–402.

Mégessier, C. (1975, March). The Ap star 108 Aqr. II - The oblique rotator model. *Astronomy and Astrophysics, 39*, 263–273.

Michaud, G., Charland, Y., & Megessier, C. (1981, November). Diffusion models for magnetic Ap-Bp stars. *Astronomy and Astrophysics, 103*, 244–262.

Mihalas, D. (1973, September) On the helium-spectrum variations of 56 Arietis and a Centauri. *Astrophysical Journal, 184*, 851–871.

Morgenthaler, A., Petit, P., Saar, S., Solanki, S. K., Morin, J., Marsden, S. C., et al. (2012, April). Long-term magnetic field monitoring of the Sun-like star ξ Bootis A. *Astronomy and Astrophysics, 540*, A138.

Morin, J., Donati, J.-F., Petit, P., Delfosse, X., Forveille, T., Albert, L., et al. (2008, September) Large-scale magnetic topologies of mid M dwarfs. *Monthly Notices of the Royal Astronomical Society, 390*, 567–581.

Nesvacil, N., Lüftinger, T., Shulyak, D., Obbrugger, M., Weiss, W., Drake, N. A., et al. (2012, January). Multi-element Doppler imaging of the CP2 star HD 3980. *Astronomy and Astrophysics, 537*, A151.

Petit, P., Dintrans, B., Solanki, S. K., Donati, J.- F., Aurière, M., Lignières, F., et al. (2008, July). Toroidal versus poloidal magnetic fields in Sun-like stars: a rotation threshold. *Monthly Notices of the Royal Astronomical Society, 388*, 80–88.

Piskunov, N., & Kochukhov, O. (2002, January). Doppler Imaging of stellar magnetic fields. I. Techniques. *Astronomy and Astrophysics, 381*, 736–756.

Piskunov, N. E., Tuominen, I., & Vilhu, O. (1990, April). Surface imaging of late-type stars. *Astronomy and Astrophysics, 230*, 363–370.

Polosukhina, N., Kurtz, D., Hack, M., North, P., Ilyin, I., Zverko, J., et al. (1999, November). Lithium on the surface of cool magnetic CP stars I. Summary of spectroscopic observations with three telescopes. *Astronomy and Astrophysics, 351*, 283–291.

Reiners, A., & Basri, G. (2007, February). The first direct measurements of surface magnetic fields on very low mass stars. *Astrophysical Journal, 656*, 1121–1135.

Reiners, A., & Basri, G. (2009, March). On the magnetic topology of partially and fully convective stars. *Astronomy and Astrophysics, 496*, 787–790.

Rice, J. B., Holmgren, D. E., & Bohlender, D. A. (2004, September) The distribution of oxygen on the surface of the Ap star θ Aur. An abundance Doppler image to compare with ϵ UMa. *Astronomy and Astrophysics, 424*, 237–244.

Rice, J. B., Strassmeier, K. G., & Kopf, M. (2011, February). The surface of V410 Tauri. *Astrophysical Journal, 728*, 69.

Rice, J. B., & Wehlau, W. H. (1991, June). The range of abundances of iron, chromium, and silicon over the surfaces of the CP stars Epsilon Ursae Majoris and Theta Aurigae. *Astronomy and Astrophysics, 246*, 195–198.

Rice, J. B., Wehlau, W. H., & Holmgren, D. E. (1997, October). The distribution of oxygen on the surface of ϵ UMa: an abundance distribution Doppler image. *Astronomy and Astrophysics, 326*, 988–994.

Rice, J. B., Wehlau, W. H., & Khokhlova, V. L. (1989, January). Mapping stellar surfaces by Doppler imaging - Technique and application. *Astronomy and Astrophysics, 208*, 179–188.

Rosén, L., & Kochukhov, O. (2012, December). How reliable is Zeeman Doppler imaging without simultaneous temperature reconstruction? *Astronomy and Astrophysics, 548*, A8.

Rosén, L., Kochukhov, O., & Wade, G. A. (2013, November). Strong variable linear polarization in the cool active star II Peg. *Monthly Notices of the Royal Astronomical Society, 436*, L10–L14.

Rosén, L., Kochukhov, O., & Wade, G. A. (2015, June). First Zeeman Doppler imaging of a cool star using all four stokes parameters. *Astrophysical Journal, 805*, 169.

Rusomarov, N., Kochukhov, O., Piskunov, N., Jeffers, S. V., Johns-Krull, C. M., Keller, C. U., et al. (2013, October). Three-dimensional magnetic and abundance mapping of the cool Ap star HD 24712 . I. Spectropolarimetric observations in all four Stokes parameters. *Astronomy and Astrophysics, 558*, A8.

Rusomarov, N., Kochukhov, O., Ryabchikova, T., & Piskunov, N. (2015). Three-dimensional magnetic and abundance mapping of the cool Ap star HD 24712 . II. Magnetic Doppler imaging in all four Stokes parameters. *Astronomy and Astrophysics, 573*, A123.

Ryabchikova, T., Piskunov, N., Kochukhov, O., Tsymbal, V., Mittermayer, P., & Weiss, W. W. (2002, March). Abundance stratification and pulsation in the atmosphere of the roAp star boldmath gamma Equulei. *Astronomy and Astrophysics, 384*, 545–553.

Ryabchikova, T. A., Malanushenko, V. P., & Adelman, S. J. (1999, November). Orbital elements and abundance analyses of the double-lined spectroscopic binary alpha Andromedae. *Astronomy and Astrophysics, 351*, 963–972.

Ryabchikova, T. A., Pavlova, V. M., Davydova, E. S., & Piskunov, N. E. (1996, November). Surface distribution of chromium on the CP2 star HD 220825 (κ Psc). *Astronomy Letters, 22*, 821–826.

Saio, H., & Gautschy, A. (2004, May). Axisymmetric p-mode pulsations of stars with dipole magnetic fields. *Monthly Notices of the Royal Astronomical Society, 350*, 485–505.

Schrijver, C. J., & Title, A. M. (2001, April). On the formation of polar spots in sun-like stars. *Astrophysical Journal, 551*, 1099–1106.

Silvester, J., Kochukhov, O., & Wade, G. A. (2014, May). Stokes IQUV magnetic Doppler imaging of Ap stars - II. Next generation magnetic Doppler imaging of α^2 CVn. *Monthly Notices of the Royal Astronomical Society, 440*, 182–192.

Silvester, J., Wade, G. A., Kochukhov, O., Bagnulo, S., Folsom, C. P., & Hanes, D. (2012, October). Stokes IQUV magnetic Doppler imaging of Ap stars - I. ESPaDOnS and NARVAL observations. *Monthly Notices of the Royal Astronomical Society, 426*, 1003–1030.

Skelly, M. B., Donati, J.-F., Bouvier, J., Grankin, K. N., Unruh, Y. C., Artemenko, S. A., et al. (2010, March). Dynamo processes in the T Tauri star V410 Tau. *Monthly Notices of the Royal Astronomical Society, 403*, 159–169.

Stibbs, D. W. N. (1950). A study of the spectrum and magnetic variable star HD 125248. *Monthly Notices of the Royal Astronomical Society, 110*, 395–404.

Strassmeier, K. G. (2009, September) Starspots. *Astronomy and Astrophysics Review, 17*, 251–308.

Strassmeier, K. G. (2011, August). The zoo of starspots. In D. Prasad Choudhary & K. G. Strassmeier (Eds.), *IAU Symposium* (Vol 273, pp. 174–180).

Strassmeier, K. G., Rice, J. B., Wehlau, W. H., Vogt, S. S., Hatzes, A. P., Tuominen, I., et al. (1991, July). Doppler imaging of high-latitude spot activity on HD 26337. *Astronomy and Astrophysics, 247*, 130–147.

Unruh, Y. C., Collier Cameron, A., & Cutispoto, G. (1995, December). The evolution of surface structures on Ab-Doradus. *Monthly Notices of the Royal Astronomical Society, 277*, 1145.

Vogt, S. S., Hatzes, A. P., Misch, A. A., & Kürster, M. (1999, April). Doppler imagery of the spotted RS Canum Venaticorum Star HR 1099 (V711 Tauri) from 1981 to 1992. *The Astrophysical Journal Supplement Series, 121*, 547–589.

Vogt, S. S., & Penrod, G. D. (1983a, December). Detection of high-order nonradial oscillations on the rapid rotator Zeta Ophiuchi and their link with Be type outbursts. *Astrophysical Journal Supplement Series, 275,* 661–682.

Vogt, S. S., & Penrod, G. D. (1983b, September) Doppler imaging of spotted stars - Application to the RS Canum Venaticorum star HR 1099. *Publications of the Astronomical Society of the Pacific, 95,* 565–576.

Vogt, S. S., Penrod, G. D., & Hatzes, A. P. (1987, October). Doppler images of rotating stars using maximum entropy image reconstruction. *Astrophysical Journal, 321,* 496–515.

Wade, G. A., Donati, J.-F., Landstreet, J. D., & Shorlin, S. L. S. (2000, April). Spectropolarimetric measurements of magnetic Ap and Bp stars in all four Stokes parameters. *Monthly Notices of the Royal Astronomical Society, 313,* 823–850.

Index

© Springer International Publishing Switzerland 2016
J.-P. Rozelot, C. Neiner (eds.), *Cartography of the Sun and the Stars*,
Lecture Notes in Physics 914, DOI 10.1007/978-3-319-24151-7

Printed in the United States
By Bookmasters